Stochasticity and Partial Order

Mathematics and Its Applications

Volume 9

Peter M. Alberti and Armin Uhlmann
Department of Physics, Karl-Marx-University, Leipzig

Stochasticity and Partial Order

Doubly Stochastic Maps and Unitary Mixing

VEB Deutscher Verlag der Wissenschaften, Berlin

D. Reidel Publishing Company
Dordrecht : Holland / Boston : U.S.A. / London : England

Library of Congress Cataloging in Publication Data

Alberti, Peter M., 1947–
Stochasticity and partial order.

(Mathematics and its applications; v. 9)
Bibliography: p.
Includes index.
1. Statistical physics. 2. Mappings (Mathematics) 3. Stochastic processes.
I. Uhlmann, Armin. II. Title. III. Series: Mathematics and its applications (D. Reidel Publishing Co.); v. 9.
QC174.8.A39 530.1'5 81-19922
ISBN 90-277-1350-2 (Reidel) AACR2

Distributors for the U.S.A. and Canada
Kluwer Boston Inc.,
190 Old Derby Street, Hingham, MA 02043, U.S.A.

Distributors for all remaining countries
Kluwer Academic Publishers Group,
P. O. Box 322, 3300 AH Dordrecht, Holland.

D. Reidel Publishing Company is a member of the Kluwer Group.

Printed in the German Democratic Republic.

Editor's Preface

Approach your problems from the right end and begin with the answers. Then, one day, perhaps you will find the final question.

'The Hermit Clad in Crane Feathers' in R. Van Gulik's *The Chinese Maze Murders*.

It isn't they can't see the solution. It is that they can't see the problem.

G. K. Chesterton, *The Scandal of Father Brown* 'The Point of a Pin'.

Growing specialization and diversification have brought a host of monographs and textbooks on increasingly specialized topics. However, the 'tree' of knowledge of mathematics and related fields does not grow only by putting forth new branches. It also happens, quite often in fact, that branches which were thought to be completely disparate are suddenly seen to be related.

Further, the kind and level of sophistication of mathematics applied in various sciences has changed drastically in recent years: measure theory is used (non-trivially) in regional and theoretical economics; algebraic geometry interacts with physics; the Minkowski lemma, coding theory and the structure of water meet one another in packing and covering theory; quantum fields, crystal defects and mathematical programming profit from homotopy theory; Lie algebras are relevant to filtering; and prediction and electrical engineering can use Stein spaces.

This series of books, *Mathematics and Its Applications*, is devoted to such (new) interrelations as *exampla gratia*:

- a central concept which plays an important role in several different mathematical and/or scientific specialized areas;
- new applications of the results and ideas from one area of scientific endeavor into another;
- influences which the results, problems and concepts of one field of enquiry have and have had on the development of another.

With books on topics such as these, of moderate length and price, which are stimulating rather than definitive, intriguing rather than encyclopaedic, we hope to contribute something towards better communication among the practitioners in diversified fields.

The present book is, in my opinion, a good example of the philosophy underlying this program. The starting point is a certain partial order relation on sequences of positive numbers with the same sum which occurs in many seemingly unrelated areas of mathematics including the representation theory of the symmetric groups, combinatorics, vectorbundles on the Riemann sphere (highest weight) representations of Lie algebras, linear system theory and thermodynamics and stochastics. The last relation named is the main topic in this book and starting with doubly stochastic matrices we meet among others entropy like quantities, W^*-algebras, irreversible processes and some noncommutative probability and ergodicity theory.

The unreasonable effectiveness of mathematics in science . . .

Eugene Wigner

Well, if you knows of a better 'ole, go to it.

Bruce Bairnsfather

What is now proved was once only imagined.

William Blake

As long as algebra and geometry proceeded along separate paths, their advance was slow and their applications limited.

But when these sciences joined company, they drew from each other fresh vitality and thenceforward marched on at a rapid pace towards perfection.

Joseph Louis Lagrange

Krimpen a/d IJssel
Sept. 1981

Michiel Hazewinkel

Contents

Introduction

It is our intention to explain and to prove certain relations between stochasticity and partial order within this book.

An elementary but typical example is the following theorem due to RADO. Let us consider two n-tuples of real numbers, $\boldsymbol{a} = \{a_1, a_2, \dots\}$ and $\boldsymbol{b} = \{b_1, b_2, \dots\}$, which are decreasingly ordered, i.e. $a_1 \geqq a_2 \geqq \cdots$ and $b_1 \geqq b_2 \geqq \cdots$ respectively, and which fulfil $a_1 + a_2 + \cdots + a_n = b_1 + b_2 + \cdots + b_n$. Then $\boldsymbol{a}$ is a convex sum of n-tuples $\boldsymbol{b}^{(k)}$, where every $\boldsymbol{b}^{(k)}$ denotes a permutation of the n-tuple $\boldsymbol{b}$, if and only if

$$\forall m\colon \sum_{j=1}^{m} a_j \leqq \sum_{j=1}^{m} b_j \,. \tag{1}$$

Using Birkhoff's result concerning the structure of doubly stochastic matrices, one at once infers the equivalence of (1) under the assumptions above with the existence of a doubly stochastic matrix T mapping $\boldsymbol{b}$ onto $\boldsymbol{a}$, i.e. $T\boldsymbol{b} = \boldsymbol{a}$.

For reasons of physical interpretation we call $\boldsymbol{a}$ "more chaotic" than $\boldsymbol{b}$ iff $\boldsymbol{a} = T\boldsymbol{b}$ with a doubly stochastic T. In the first chapter, there are several examples explaining this point. In particular we refer to the increase of all "entropy-like quantities" if probability vectors become more chaotic.

After mainly dealing with now-classical results and some of their recent applications in Chapter 1, we extend these considerations to matrices in Chapter 2. Though there is little known of the structure of doubly stochastic transformations in matrix spaces (see Definition 2-2), one can establish similar connections between such doubly stochastic and related maps and appropriate partial (pre-)orderings. In the first two chapters we select, hopefully, a representative part of our present knowledge in that domain, although several branches cannot even be touched.

In Chapter 3 we indicate how to relate all that to the geometry of state spaces of finite-dimensional $*$-algebras and intend to convince the reader of the inescapability of extending it to general W^*-algebras.

The main tools and proofs for doing so are given in Chapters 4 and 5. We believe part of our techniques to be of general interest in the theory of W^*-algebras; for example, the Ky Fan functionals, the c-ideal, the central-valued convex trace, and the Σ-property.

In the last chapter, we introduce the dual structure to the partial (pre-)ordering $\succ$ of the state space. Let a, b be two Hermitian elements of a W^*-algebra. We write $a \succ b$ iff for all positive linear functionals ω of the algebra, we have

$$\sup \omega(a^u) \geqq \sup \omega(b^u) \,. \tag{2}$$

Here a^u, b^u denote unitary transforms of the elements a, b, and the supremum has to run through all unitaries of the algebra in question. In this way one gets an extension of the famous von Neumann-Murray classification of pairs of projectors to pairs of Hermitian elements.

We do not present analogues in W^*-algebras for all the theorems of the first two chapters. However, using the technicalities developed, the skilled reader can find many of them by himself. Furthermore, one may obviously extend this theory completely to C^*-algebras with "sufficiently many projections", i.e. to AW^*-algebras. For general C^*-algebras, however, the results are essentially incomplete.

As already mentioned, there are parts of our work that are related to physics, including the description of some irreversible processes, the classification of mixed (in the sense of GIBBS and VON NEUMANN) states, and the problem of general diffusions. There are connections to non-commutative probability theory and non-commutative ergodicity.

Last not least, we point to the following statement: With the exception of the first chapter, the theory presented relies on the use of the unitary transformations of a W^*-algebra. Therefore, without some care, one gets trivialities for commutative algebras. In 1.9. a relevant idea for the commutative case is explained. Its implications for general commutative C^*-algebras will be considered elsewhere.*)

*) P. M. ALBERTI and A. UHLMANN, Dissipative motion in state spaces, Teubner-Texte zur Mathematik Bd. 33, Leipzig 1981.

1. Some classical results

1.1. Elementary notations

Let L be a linear space over the real numbers. A subset K of L is called *convex* iff for all $0 < t < 1$ and for all $\boldsymbol{a}$, $\boldsymbol{b} \in K$ one has $(1 - t)\,\boldsymbol{a} + t\boldsymbol{b} \in K$.

Every complex linear space can be considered as a real one. This will always be done in connection with convexity, i.e. with the definition of convex subsets.

Let K be a convex subset of a linear space L. A function (sometimes also called "a functional") f from K into the real numbers is called *convex* if on its domain of definition always

$$f((1 - t)\,\boldsymbol{a} + t\boldsymbol{b}) \leqq (1 - t)\,f(\boldsymbol{a}) + tf(\boldsymbol{b})\,.$$

If f is convex, the function $-f$ is, by definition, *concave*. A function which is both, convex and concave, is called *affine*. If a function depends on two elements of a convex set K, i.e.

$$f\colon \boldsymbol{a}, \boldsymbol{b} \to f(\boldsymbol{a}, \boldsymbol{b})\,,$$

it is said to be *simultaneously convex* iff it is convex as a function

$$\boldsymbol{a} + \boldsymbol{b} \to f(\boldsymbol{a}, \boldsymbol{b})$$

on the subset $K + K$ of the direct sum $L + L$. Similar conventions are valid for functions of several variables.

The supremum of a family of convex functions is convex again and the infimum of every family of concave functions is concave, too. Hence, if L carries a locally convex topological structure, one is advised to consider lower semicontinuous (l.s.c.) convex and upper semicontinuous (u.s.c.) concave functions. In particular, every l.s.c. convex function on a compact convex set is the supremum of an appropriate family of continuous affine functionals. For more refined results see [15, 58].

1.2. The linear space l^1 in n dimensions

Throughout Chapter 1 we denote by $\boldsymbol{l}^1$ the real linear space of n-tuples

$$\boldsymbol{a} = \{a_1, a_2, \dots, a_n\}\,, \qquad a_j \in \mathbf{R}\,, \tag{1}$$

equipped with the $\boldsymbol{l}^1$-norm

$$\|\boldsymbol{a}\|_1 := |a_1| + |a_2| + \cdots + |a_n|\,. \tag{2}$$

The number a_i are sometimes called *components* of $\boldsymbol{a}$.

An element $\boldsymbol{a}$ of l^1 is said to be *positive*, and we write

$$\boldsymbol{a} \geqq 0$$

in that case, iff all its components are non-negative real numbers. The stronger condition $\forall j \colon a_j > 0$ will be referred to as *strictly positive*.

The notation of positivity is connected with a partial ordering in l^1: One writes

$$\boldsymbol{a} \geqq \boldsymbol{b} \quad \text{iff} \quad \boldsymbol{a} - \boldsymbol{b} \geqq 0$$

Let us now consider a linear map T from l^1 into itself,

$$T \colon l^1 \to l^1 . \tag{3}$$

Such a map can be characterized uniquely by a matrix (t_{ik}), where the indices i, k run from 1 to n, which gives, for an arbitrary element (1), the jth component $(T\boldsymbol{a})_j$ of $T\boldsymbol{a}$ by

$$(T\boldsymbol{a})_j = \sum_{i=1}^{n} t_{ji} a_i . \tag{4}$$

To indicate this connection we occasionally identify T with its matrix (t_{jk}) and write $T = (t_{jk})$. As usual, the norm $||T||$ of a map (3) is defined by

$$||T||_1 := \sup ||T\boldsymbol{a}||_1/||\boldsymbol{a}||_1 \tag{5}$$

where $\boldsymbol{a}$ runs through all the elements of l^1 with the exception of the zero of l^1.

A map, the norm of which is smaller than or equal to 1, is called a *contraction* or *contracting*. An *isometry* T is a map preserving the norm $||.||_1$, i.e. for all $\boldsymbol{a} \in l^1$ the equation $||T\boldsymbol{a}||_1 = ||\boldsymbol{a}||_1$ holds. The finite-dimensionality of l^1 implies the invertibility of every isometry. Hence every isometry is a one-to-one map from l^1 onto l^1.

The norm of a map can easily be calculated from the matrix representation (4). Obviously, from $\forall i \colon \sum_j |t_{ij}| \leqq 1$ we get, for the map $T = (t_{ij})$, the estimation $||T|| \leqq 1$, for (4) implies

$$||T\boldsymbol{a}||_1 \leqq \sum_i \sum_j |t_{ij}| \, |a_i| .$$

On the other hand, the mentioned condition is necessary for $||T|| \leqq 1$, too. To see this, one chooses the element $\boldsymbol{a}_i$ of l^1 having all its components zero except its ith component which is chosen to be 1. Then $||\boldsymbol{a}_i||_1 = 1$ and

$$||T\boldsymbol{a}_i||_1 = \sum_j |t_{ji}| \leqq ||T|| .$$

Hence, the desired result reads

$$||T|| = \sup_i \sum_j |t_{ji}| \quad \text{for} \quad T = (t_{ji}) . \tag{6}$$

Lemma 1-1. *Let T be an isometry. Then there is a permutation*

$$j \to i_j , \qquad j = 1, \ldots , n ,$$

of the integers 1, 2, ... , n *and* n *choices of* ε_j *among* $+1$ *and* -1, *i.e.* $\varepsilon_j = \pm 1$, *such that for all* $\boldsymbol{a}$

$$T\{a_1, a_2, \dots , a_n\} = \{\varepsilon_1 a_{j_1}, \varepsilon_2 a_{j_2}, \dots , \varepsilon_n a_{j_n}\} . \tag{7}$$

Proof. T being invertible, T^{-1} is also an isometry. Hence, the unit ball

$$\boldsymbol{E} := \{\boldsymbol{a} \in \boldsymbol{l}^1 : ||\boldsymbol{a}||_1 \leqq 1\}$$

is mapped one-to-one onto itself under the action of T. Thus, T being a convex automorphism of the convex set $\boldsymbol{E}$, it converts every extremal point of the convex set $\boldsymbol{E}$ into an extremal point. An extremal point of $\boldsymbol{E}$ is characterized by having only one non-vanishing component which equals either $+1$ or -1. Especially with $\boldsymbol{a}_i = (0, \dots , 1, \dots , 0)$, the $+1$ being in the ith position, we get $T\boldsymbol{a}_i = \varepsilon_i \boldsymbol{a}_{j_i}$ with $\varepsilon_i = \pm 1$ and $i \to j_i$ defines a permutation. ∎

The lemma tells us the structure of the matrix $(t_{ik}) = T$ of an isometry T: In every row and in every column there is one and only one non-vanishing element which is either $+1$ or -1. If this non-vanishing element always equals $+1$, we call T a *permutation map*.

By calling a map T *positivity preserving* if $\boldsymbol{a} \geqq 0$ always implies $T\boldsymbol{a} \geqq 0$, we arrive at the maps $T = (t_{ik})$ with $\forall\ i, k$: $t_{ik} \geqq 0$. They play an important role.

As a consequence of Lemma 1-1 we may state:

Every positivity preserving isometry is a permutation map and vice versa.

1.3. Stochastic and doubly stochastic maps

Definition 1-1. A map from $\boldsymbol{l}^1$ into $\boldsymbol{l}^1$ is said to be *stochastic* if it is positivity preserving and satisfies

$$||T\boldsymbol{a}||_1 = ||\boldsymbol{a}||_1 \quad \text{for all} \quad \boldsymbol{a} \geqq 0 . \tag{8}$$

A map T from $\boldsymbol{l}^1$ into $\boldsymbol{l}^1$ is said to be *doubly stochastic* if it is a convex linear combination of positivity preserving isometries.

In the definition above, one should be reminded of the fact that the positivity preserving isometries are just the permutation maps.

Before studying the stochastic and doubly stochastic maps, we show a simple lemma:

Lemma 1-2. *Every doubly stochastic map is stochastic.*

The *proof* runs as follows. Because every convex combination of positivity preserving maps is positivity preserving again, we have only to show that every doubly stochastic map T acts isometrically on the cone

$$\{\boldsymbol{a} \in \boldsymbol{l}^1 : \boldsymbol{a} \geqq 0\} . \tag{9}$$

The above definition provides us with a representation

$$T = t_1 P_1 + t_2 P_2 + \dots + t_m P_m , \tag{10}$$

$$t_1 + t_2 + \dots + t_m = 1 \quad \text{and} \quad \forall j: t_j \geqq 0 \tag{10a}$$

where $P_1, P_2, \ldots$ are assumed to be permutation maps. Using (2) we see $\|\boldsymbol{a} + \boldsymbol{b}\|_1 = \|\boldsymbol{a}\|_1 + \|\boldsymbol{b}\|_1$ within the cone (9). By (10) we have therefore

$$\|T\boldsymbol{a}\|_1 = \sum t_j \|P_j \boldsymbol{a}\|_1$$

for $\boldsymbol{a} \geqq 0$. The desired equality (8) now follows from (10a), because the maps P_i are isometries. ▮

Lemma 1-3. *Every stochastic map is of norm one.*

Proof. Using (8) we find it necessary only to show contractivity for every stochastic map T. To do this we decompose every element $\boldsymbol{a} = \{a_1, a_2, \ldots, a_n\}$ into its positive and negative part $\boldsymbol{a}'$ and $\boldsymbol{a}''$. It is $\boldsymbol{a}'' = \boldsymbol{a}' - \boldsymbol{a}$ and the components a_j' of $\boldsymbol{a}'$ are given by

$$a_i' = a_i \quad \text{if} \quad a_i \geqq 0 \quad \text{and} \quad a_i' = 0 \quad \text{otherwise}.$$

$\boldsymbol{a}'$ and $\boldsymbol{a}''$ both are positive. Thus (8) yields

$$\|T\boldsymbol{a}\|_1 = \|T(\boldsymbol{a}_1' - \boldsymbol{a}'')\|_1 \leqq \|T\boldsymbol{a}'\|_1 + \|T\boldsymbol{a}''\|_1 = \|\boldsymbol{a}'\|_1 + \|\boldsymbol{a}''\|_1$$

and the last the right-hand expression equals $\|\boldsymbol{a}\|_1$. ▮

Theorem 1-4. *The map $T = (t_{ik})$ is a stochastic one if and only if*

$$\forall\, i, k\colon t_{ik} \geqq 0 \qquad \textit{and} \qquad \forall\, k\colon \sum_i t_{ik} = 1\,. \tag{11}$$

Remark. A matrix satisfying (11) is called a *stochastic matrix.*

Proof. Either from (11) or from Definition 1-1 we have $t_{ik} \geqq 0$ for all i, k. Thus for $\boldsymbol{a} \geqq 0$ we get, denoting its components by a_k,

$$\|T\boldsymbol{a}\|_1 = \sum_k a_k \sum_i t_{ik}\,.$$

Now, (11) immediately implies (8). On the other hand, we know $\|T\| = 1$ by Lemma 1-3. Therefore, by (6) we have

$$\sum_k a_k \sum_i t_{ik} \leqq \sum_k a_k = \|\boldsymbol{a}\|_1$$

and the equality sign required by (8) cannot hold if $\sum_i t_{ik} < 1$ for one k. ▮

From now on, it is necessary and convenient to make explicit use of the following: The linear maps from $\boldsymbol{l}^1$ into $\boldsymbol{l}^1$ form a linear space in a natural way. This linear structure is defined by

$$(T + T')\,\boldsymbol{a} := T\boldsymbol{a} + T'\boldsymbol{a} \qquad \text{for all} \qquad \boldsymbol{a} \in \boldsymbol{l}^1$$

and

$$(\lambda T)\,\boldsymbol{a} = \lambda(T\boldsymbol{a}) \qquad \text{for} \qquad \lambda \in \mathbf{R}\,, \qquad \boldsymbol{a} \in \boldsymbol{l}^1\,.$$

The notation "convex set of maps" is to be understood as being relative to this linear structure.

Let us now introduce the notation

$$\mathrm{ST}(\boldsymbol{l}^1) \qquad \text{or, simply,} \qquad \mathrm{ST}$$

for the set of all stochastic maps.

Lemma 1-5. *The set*

$$\text{ST} := \textit{set of all stochastic maps}$$

is a compact convex subset of the linear space of all maps from l^1 into l^1.

Indeed, (11) identifies ST as a bounded closed set of a finite-dimensional linear space. Further, with T, T' stochastic, we find $\frac{1}{2}T + \frac{1}{2}T'$ positivity preserving and this implies for all $\boldsymbol{a} \geqq 0$

$$\|\tfrac{1}{2}\, T\boldsymbol{a} + \tfrac{1}{2}\, T'\boldsymbol{a}\|_1 = \tfrac{1}{2}\, \|T\boldsymbol{a}\|_1 + \tfrac{1}{2}\, \|T'\boldsymbol{a}\|_1 \,.$$

Because of (8) the latter expression equals $\|\boldsymbol{a}\|_1$. ∎

Being compact, convex, and finite-dimensional, every element of ST can be decomposed into a convex linear sum of its extremal elements (Minkowski).

Let us now assume, $T = (t_{ik})$ is extremal in ST. We shall show $t_{jk} = 1$ whenever $t_{jk} \neq 0$. To this end we firstly consider what happens if $0 < t_{11} < 1$. We construct T' by setting $t'_{11} = 1$, $t'_{1k} = 0$ if $k \neq 1$, and $t'_{ik} = t_{ik}$ for $i \neq 1$. Clearly, $T' = (t'_{ik})$ is stochastic and $T - t_{11}T'$ is positivity preserving. Introducing T'' by the equation

$$T = t_{11}T' + (1 - t_{11})\, T'',$$

we arrive at a contradiction: T'' is stochastic (positivity preserving is trivial, the sum of every column is equal to 1 because this is so with T and T') and is different from T. Hence, $t_{11} \neq 0$ implies $t_{11} = 1$. Obviously, our argument does not depend on the special position of the matrix element in the matrix (t_{ik}). Therefore, an extremal element of ST can have zero's and one's in its matrix representation only. On the other hand, because every matrix element of a stochastic map is a number between zero and one, it is a simple exercise to see that the derived property of an extremal stochastic map is also to be a sufficient condition for extremality in ST. This proves:

Theorem 1-6. *A stochastic map $T = (t_{ik})$ is extremal in the convex set* ST *if and only if its matrix representation consists of zero's and one's only.*

A map $T = (t_{ik})$ is extremal stochastic iff in each of its columns there is one and only one element different from zero and this non-zero element has to be one.

The last assertion comes from the preceding one and Theorem 1-4.

Remark. In the proof of Theorem 1-6 the finite-dimensionality of l^1 has been irrelevant.

A further basic property of stochastic matrices directly follows from the Definition 1-1: The composition T_1T_2 of two stochastic maps is stochastic, too. Further, let T and T^{-1} both be stochastic maps. (In writing T^{-1} we assume its existence.) Then T has to be an isometry by Lemma 1-3 and even a permutation map as an already mentioned consequence of Lemma 1-1.

This way we have seen the following statements to be equivalent:

(a) T is a permutation map,

(b) T is a positivity preserving isometry,

(c) T is stochastic, T^{-1} exists and is stochastic, too,

(d) T is extremal in ST, and T^{-1} exists.

The last statement is a corollary to Theorem 1-6: If the extremal $T = (t_{ik})$ is not a permutation, then there is one row (at least) which does not contain exactly one $+1$. Because the total sum of all matrix elements equals n, there had to be a row consisting of zeros only, and T cannot have an inverse. ∎

In the following we prepare for the study of the doubly stochastic matrices. At first we define certain convex subsets of ST. Denoting by $\boldsymbol{b}'$, $\boldsymbol{b}''$ two elements of $\boldsymbol{l}^1$, we introduce the convex compact set

$$\mathrm{ST}(\boldsymbol{b}', \boldsymbol{b}'') = \{T \in \mathrm{ST}: T\boldsymbol{b}' = \boldsymbol{b}''\} . \tag{12}$$

This set has a special property:

Assume T and T' to be elements of $\mathrm{ST}(\boldsymbol{b}', \boldsymbol{b}'')$, and t a real number between zero and one. If now the map

$$(T - tT')(1 - t)^{-1} = T''$$

is positivity preserving, then $T'' \in \mathrm{ST}(\boldsymbol{b}', \boldsymbol{b}'')$ and we have the decomposition

$$T = (1 - t)\, T'' + tT' .$$

Therefore, if T'' is positivity preserving, T cannot be an extremal element of the convex set $\mathrm{ST}(\boldsymbol{b}', \boldsymbol{b}'')$.

In passing, we remark that the just-mentioned property is shared by intersections of an arbitrary number of sets of type (12).

Lemma 1-7. *Let $T = (t_{ik})$ be an extremal and $T' = (t'_{ik})$ an arbitrary element of $\mathrm{ST}(\boldsymbol{b}', \boldsymbol{b}'')$. Assume that for every pair j, s of indices for which $t_{js} = 0$ we also have $t'_{js} = 0$. Then $T = T'$.*

As a consequence, the extremal elements of (12) are uniquely determined by their distribution of zeros in the matrix representation.

Proof. The argument is essentially connected with the finite dimensionality of our $\boldsymbol{l}^1$. Namely, under the hypothesis of the lemma, one can find a small positive real number t such that $T - tT'$ is positivity preserving. We now divide by $(1 - t)$ and use the special property of (12) discussed above. ∎

In [12] the extremal elements of (12) in the general case are described. Here we restrict ourselves to the special $\boldsymbol{b}' = \boldsymbol{b}'' = \boldsymbol{e}$, where

$$\boldsymbol{e} = (1/n)\{1, 1, \ldots, 1\} .$$

Having in mind Definition 1-1, the following theorem is nothing less than a famous statement due to Birkhoff.

Theorem 1-8. *The following three properties are equivalent one-to-another:*

(a) *T is doubly stochastic,*

(b) $T \in \mathrm{ST}(\boldsymbol{e}, \boldsymbol{e})$,

(c) *T is stochastic and its matrix representation $T = (t_{ik})$ satisfies*

$$\forall\, i: \quad \sum_k t_{ik} = 1 . \tag{13}$$

Proof. Every permutation map has in each of its rows exactly one element different from zero and this matrix element is of value $+1$. Thus (13) is valid for permutation maps and therefore, according to Definition 1-1, valid for every doubly stochastic map matrix. Hence (c) follows from (a). But (b) easily follows from (c). Minkowski's theorem allows a representation of every element of ST(e, e) as a convex linear combination of the extremal points of ST(e, e). Now the proof is reduced to the next lemma. ▮

Lemma 1-9. *T is extremal in* ST(e, e) *iff T is a permutation map.*

Proof. If T is a permutation map then it is extremal even in ST and, consequently, in ST(e, e). Let us now assume T to be extremal in ST(e, e), and let us denote its matrix representation by (t_{ik}). If we could find a permutation $i \to j_i$ such that all the numbers $t_{1i_1}, t_{2i_2}, \ldots$ are different from zero, we are done. For constructing $T' = (t'_{ik})$ with $t_{ij_i} = 1$ and $t_{ik} = 0$ otherwise, we are allowed to apply Lemma 1-7. The result is $T = T'$ and T' is a permutation map by construction.

Now we have to look for the existence of the desired permutation. Having in mind that (b) implies (c) in Theorem 1-8, as has already been shown, we need to confirm the following:

Let (t_{ik}) be a matrix with non-negative matrix elements only for which the sum of every row and of every column equals one. Then there is a permutation

$$i \to i_j: \prod_i t_{ii_j} \neq 0 . \qquad (*)$$

At the end of the proof we shall use the invariance of this property againts permutations between rows and between columns.

Let s and r now denote the smallest naturals with

$$t_{ik} = 0 \qquad \text{for } i > s \text{ and } k > r .$$

According to this choice, in every row and every column of the submatrix (t_{ik}, $i \leqq s$, $k \leqq r$) there is a non-zero element. Hence, the sum of the last $n - s$ elements of one of the first r columns is strictly less than one. On the other hand, the sum of the first r elements of the last $n - s$ columns is one. Thus the sum of the first r columns is larger than $n - s$:

$$r > n - s . \qquad (**)$$

There is a pure combinatorical lemma due to Frobenius and Schur according to which (*) can either be fulfilled or a submatrix consisting of zeros only and with

$$(\text{sum of columns}) + (\text{sum of rows}) \geqq n + 1$$

can be constructed by respectively permuting rows and columns only. But

$$(n - s) + (n - r) \geqq n + 1$$

contradicts (**). Hence (*) can be fulfilled (see [55]). ▮

Remarks.

(a) The problem, to list the extremal elements for a general set ST(b', b''), was solved recently (see [12]). As a corollary one gets a further proof of Lemma 1-9.

(b) The core of the Birkhoff theorem is in proving that there are no more extremal elements in ST(e, e) than permutations.

(c) A stochastic matrix, satisfying (13), is called *doubly stochastic.*

1.4. The relation $\succ$

We now use the concept of doubly stochastic maps to introduce a partial pre-ordering in l^1 (see [34, 38, 48, 53, 55, 72, 94, 98]).

Definition 1-2. For two elements $\boldsymbol{a}$, $\boldsymbol{b}$ of l^1 we write

$$\boldsymbol{b} \succ \boldsymbol{a}$$

and call $\boldsymbol{b}$ *more mixed* or (equivalently) *more chaotic than* $\boldsymbol{a}$ if and only if there is a doubly stochastic map T with

$$\boldsymbol{b} = T\boldsymbol{a} .$$

We first have to comment on this terminology. In statistical physics, one may describe an important class of irreversible phenomena by this relation (see 1.8. for some indications), i.e. by applying doubly stochastic maps on probability distributions characterizing states of a physical system in the sense of Gibbs (and Boltzmann). One should think of the total set of such states as a convex set, the extremal elements of which are associated with "pure states" (for example, points of the phase space in mechanics). Coming from the extreme boundary of this set more and more into its interior, we lose correlations and structure, and arrive at more "disorder", a more and more chaotic internal movement of the system's constituents. In a rough heuristic sense, $\succ$ shows the "direction of irreversibility" from the pure states, where probability becomes certainty.

Thirring [89] therefore also has used the term "less pure" for "more chaotic". The term "more mixed" is somewhat nicer than "more chaotic", for it correctly recalls its origin from the theory of Gibbsian mixtures. We prefer to say "more chaotic" in the following, so that we avoid misunderstanding the use of the word "mixing" in ergodic theory. There is a further comment. The preordering we are going to examine (and further similar pre-orderings) is sometimes introduced into the mathematical literature, but with the symbol $\succ$ showing exactly in the opposite direction. Hence, some caution is necessary!

Let us write

$$\boldsymbol{a} \sim \boldsymbol{b} \quad \text{iff} \quad \boldsymbol{a} \succ \boldsymbol{b} \quad \text{and} \quad \boldsymbol{b} \succ \boldsymbol{a} \tag{14}$$

and prove

Lemma 1-10. *$\boldsymbol{a} \succ \boldsymbol{b}$ and $\boldsymbol{b} \succ \boldsymbol{c}$ implies $\boldsymbol{a} \succ \boldsymbol{c}$. We have $\boldsymbol{a}' \sim \boldsymbol{a}''$ iff there is a permutation map P with $P\boldsymbol{a}' = \boldsymbol{a}''$.*

The assertion is trivial besides the "only if" part in the second statement. The set

$$\{\boldsymbol{a}: \boldsymbol{a} \succ \boldsymbol{a}'\} \tag{*}$$

is compact and convex and, according to Definition 1-1, generated by the set $\{P\boldsymbol{a}' : P \text{ is a permutation}\}$. This set therefore contains all the extremal points of (*). Every permutation P maps (*) one-to-one onto itself. Hence, every $P\boldsymbol{a}'$ is extremal in (*). But from $\boldsymbol{a}' \sim \boldsymbol{a}''$ it follows

$$\{\boldsymbol{a}: \boldsymbol{a} \succ \boldsymbol{a}'\} = \{\boldsymbol{a}: \boldsymbol{a} \succ \boldsymbol{a}''\}$$

and these two sets have to have the same extremal points which are of the form $P\boldsymbol{a}'$ in the first and $P\boldsymbol{a}''$ in the second case. ▮

Lemma 1-10 shows $\succ$ defining a partial pre-order in l^1 which, however, is not compatible with the linear structure of l^1 in the usual sense.

Lemma 1-11. *For a linear map T the relation*

$$\forall \boldsymbol{a} \in l^1: \quad T\boldsymbol{a} \succ \boldsymbol{a}$$

implies the double stochasticity of T.

The *proof* is almost trivial: For positive $\boldsymbol{a}$ we conclude $T\boldsymbol{a} \geqq 0$ and $\|\boldsymbol{a}\|_1 = \|T\boldsymbol{a}\|_1$. Thus T is stochastic. $\boldsymbol{b} \succ \boldsymbol{e}$ obviously implies $\boldsymbol{b} = \boldsymbol{e}$ by Theorem 1-8 and Definition 1-2. The same theorem now proves the assertion. ∎

Calling an element $\boldsymbol{b} \in l^1$ *maximally chaotic* if $\boldsymbol{a} \succ \boldsymbol{b}$ always implies $\boldsymbol{a} \sim \boldsymbol{b}$, we similar conclude:

Every maximally chaotic element $\boldsymbol{b}$ of l^1 is a multiple of $\boldsymbol{e}$. Otherwise there would be a permutation map P with $P\boldsymbol{b} \neq \boldsymbol{b}$ and $(1/2)(\boldsymbol{b} + P\boldsymbol{b}) = \boldsymbol{a} \prec \boldsymbol{b}$ but $\boldsymbol{a} \sim \boldsymbol{b}$ could not hold. (One may choose for P a transposition!) ∎

Define now for every real number $0 \leqq s \leqq n$

$$e_s(\boldsymbol{a}) := \sup t_1 a_1 + \cdots + t_n a_n \tag{15}$$

where $a_1, a_2, \ldots$ stands for the component of $\boldsymbol{a}$ and the supremum has to run through all n-tupels $t_1, \ldots, t_n$ with

$$\forall j: 0 \leqq t_j \leqq 1 \qquad \text{and} \qquad t_1 + t_2 + \cdots + t_n = s\,. \tag{15a}$$

Being the supremum of linear forms,

$$\boldsymbol{a} \to e_s(\boldsymbol{a})$$

is always convex. Obviously

$$e_0(\boldsymbol{a}) = 0 \qquad \text{and} \qquad e_s(\boldsymbol{a}) = a_1 + \cdots + a_n \qquad \text{for} \qquad s = n\,. \tag{16}$$

It is convenient to define $e_s = e_n$ for $s > n$.

Lemma 1-12. *Let k be an integer and $k \leqq s \leqq k + 1$. Then*

$$e_s(\boldsymbol{a}) = (1 - t)\, e_k(\boldsymbol{a}) + t e_{k+1}(\boldsymbol{a})\,; \qquad t = s - k\,. \tag{17}$$

If we have $a_1 \geqq a_2 \geqq \cdots \geqq a_n$, it follows

$$e_k(\boldsymbol{a}) = a_1 + a_2 + \cdots + a_k\,. \tag{18}$$

The simple *proof* runs as follows. Note at first

$$\forall \text{ permutations } P: \; e_s(\boldsymbol{a}) = e_s(P\boldsymbol{a})\,. \tag{19}$$

Therefore we may restrict ourselves to the situation $a_1 \geqq a_2 \geqq \cdots$. Then (18) immediately follows from the conditions (15a).

Next we note (under the decreasing order hypothesis) $a_{k+1} = e_{k+1} - e_k$ and $e_s = e_k + t a_{k+1}$. This can be rewritten to give (17). ∎

Using the convexity in $\boldsymbol{a}$ of $e_s(\boldsymbol{a})$ and (19) one concludes from Definition 1-1 for doubly stochastic maps $e_s(T\boldsymbol{a}) \leqq e_s(\boldsymbol{a})$. One part of the following theorem is proved, therefore.

Theorem 1-13. $\boldsymbol{a} \succ \boldsymbol{b}$ *if and only if*

$$e_k(\boldsymbol{a}) \leqq e_k(\boldsymbol{b}) , \qquad k = 1, 2, \ldots , n-1 , \tag{20}$$

and

$$e_n(\boldsymbol{a}) = e_n(\boldsymbol{b}) \tag{21}$$

is valid.

From $\boldsymbol{a} = T\boldsymbol{b}$ with doubly stochastic T and (13) follows (21). The inequality (20) has already been shown to follow from $\boldsymbol{a} \succ \boldsymbol{b}$. The proof that (20) together with (21) implies $\boldsymbol{a} \succ \boldsymbol{b}$ will be given after Lemma 1.15. ▮

The theorem is due to Rado [34, 48].

By (17) and $0 \leqq t \leqq 1$ it follows from (20) and (21) that

$$\boldsymbol{a} \succ \boldsymbol{b} \quad \text{implies} \quad \forall s: e_s(\boldsymbol{a}) \leqq e_s(\boldsymbol{b}) . \tag{22}$$

One may ask whether there is a variant of Theorem 1-13 using the inequalities (22) instead of (20) and (21). Now, if $\boldsymbol{a} \geqq 0$ one easily finds $e_s(\boldsymbol{a}) \geqq e_s(T\boldsymbol{a})$ for every isometry T. Together with the convexity of e_s this is already one direction in the proof of a theorem due to Markus [55].

Theorem 1-14. *Assuming* $\boldsymbol{a} \geqq 0$ *and* $\boldsymbol{b} \geqq 0$ *we have* $e_s(\boldsymbol{a}) \leqq e_s(\boldsymbol{b})$ *for all* $s \geqq 0$ *if and only if* $\boldsymbol{a}$ *is in the convex hull of the set* $\{T\boldsymbol{b};\ T \text{ isometry of } \boldsymbol{l}^1\}$.

To complete the proofs of the two last theorems we introduce some helpful notations. The map T is said to be a *k-map* if it affects at most the first k components, i.e. $\forall \boldsymbol{a}: (T\boldsymbol{a})_j = a_j$ for $j > k$. We further use

$$\boldsymbol{a} \underset{k}{\succ} \boldsymbol{b} \quad \text{iff} \quad e_i(\boldsymbol{a}) \leqq e_i(\boldsymbol{b}) \quad \text{for all} \quad i \leqq k$$

and work in the set of monotonously decreasing vectors

$$\boldsymbol{ml}^1 = \{\boldsymbol{a} = \{a_1, a_2, \ldots , a_n\} : a_1 \geqq a_2 \geqq \cdots \geqq a_n\} .$$

If $\boldsymbol{a}$ and $\boldsymbol{b}$ are contained in $\boldsymbol{ml}^1$, we write

$$\boldsymbol{a} \underset{k}{\geqq} \boldsymbol{b} \quad \text{iff} \quad a_i \geqq b_i \quad \text{for} \quad i \geqq k .$$

Generally we use the notation

$$\boldsymbol{a} \underset{k}{\geqq} \boldsymbol{b} \quad \text{iff} \quad P\boldsymbol{a} \underset{k}{\geqq} P'\boldsymbol{b} \quad \text{with} \quad P\boldsymbol{a}, P'\boldsymbol{b} \in \boldsymbol{ml}^1$$

and suitable permutation maps P, P'.

Lemma 1-15. *Let* $k \leqq n$ *and* $\boldsymbol{a}, \boldsymbol{b} \in \boldsymbol{ml}^1$. *If* $\boldsymbol{a} \underset{k}{\succ} \boldsymbol{b}$ *there exists a doubly stochastic k-map* T *with* $\boldsymbol{a} \underset{k}{\leqq} T\boldsymbol{b}$ *and* $T\boldsymbol{b} \in \boldsymbol{ml}$.

The *proof* of this lemma is by induction. If $k = 1$ the assertion is trivial. Assume now its truth for an arbitrary $k < n$. Choose the k-map T_k indicated by the lemma. Then with $\boldsymbol{c} = T_k\boldsymbol{b}$ we conclude from $\boldsymbol{a} \underset{k+1}{\succ} \boldsymbol{b}$ the relations $\boldsymbol{a} \leqq \boldsymbol{c} \in \boldsymbol{ml}^1$ (by induction assumption) and $e_{k+1}(\boldsymbol{a}) \leqq e_{k+1}(\boldsymbol{c})$. The latter inequality follows from the doubly stochasticity of the k-map T, which implies $e_i(\boldsymbol{b}) = e_i(T\boldsymbol{b})$ for all $i \geqq k$. Thus we have

(i) $\qquad a_j \leqq c_j \quad \text{for} \quad j \leqq k \quad \text{and} \quad a_1 + a_2 + \cdots + a_{k+1} \leqq c_1 + \cdots + c_{k+1} .$

If $a_{k+1} \leqq c_{k+1}$ we are done. Otherwise there is a real t with $0 < t < 1$ with

(ii) $\quad t[(c_1 - a_1) + \cdots + (c_k - a_k)] = a_{k+1} - c_{k+1} > 0 .$

Defining now

$$d_j = c_j - t(c_j - a_j) \quad \text{for} \quad j \leqq k;$$
$$d_{k+1} = a_{k+1}; \qquad d_i = c_i \quad \text{if} \quad i \geqq k+1$$

we note $\boldsymbol{d} \in \boldsymbol{ml}^1$, which follows from (i) and $a_{k+1} > c_{k+1}$. $\boldsymbol{a} \underset{k+1}{\geqq} \boldsymbol{d}$ is obvious and we only have to construct a suitable $(k+1)$-map T_{k+1}. Using the ansatz $T_{k+1} = T' T_k$, we require $T'\boldsymbol{c} = \boldsymbol{d}$. Define for fixed j, $j = 1, \ldots, k$, the map S_j by

$$(S_j\boldsymbol{x})_i = x_i \quad \text{for} \quad i \neq j \quad \text{and} \quad i \neq k+1 ,$$
$$(S_j\boldsymbol{x})_j = (1 - s_j)\, x_j + s_j x_{k+1} ,$$
$$(S_j\boldsymbol{x})_{k+1} = s_j x_j + (1 - s_j)\, x_{k+1} .$$

A map of that kind, i.e. which changes at most two components, is called *elementary*. Fixing s_j by

$$s_j = t(c_j - a_j)\,(c_j - c_{k+1})^{-1}$$

we conclude $0 \leqq s_j \leqq 1$ by (i) and (ii) and this implies double stochasticity for S_j. We shall now see that with

(iii) $\quad T' = S_1 S_2 \cdots S_k$

the induction proof comes to its end. T' is a $(k+1)$-map, it is doubly stochastic, and $(T'\boldsymbol{c})_i = d_i$ for all i. The last assertion is simple for $i \neq k$ because such an index will be changed by at most one of the maps S_j. Double stochasticity implies $\sum (T'\boldsymbol{c})_i = \sum d_i$ and this (or a small calculation) gives the required equality also in the case $i = k$. ∎

Corollary. *The k-map T in Lemma* 1-15 *can be chosen to be a product of doubly stochastic elementary maps.*

The proof of Theorems 13 and 14 can now be finished simply. First, by a permutation invariance of the assertions, we are allowed to assume $\boldsymbol{a}, \boldsymbol{b} \in \boldsymbol{ml}^1$. Applying the lemma with $k = n$, (22) implies the existence of a doubly stochastic map T with $T\boldsymbol{b} \geqq \boldsymbol{a}$. From (21) we conclude the validity of the equality sign. This proves Theorem 13. The same trick reduces the proof of Theorem 14 to the case $\boldsymbol{a} \leq \boldsymbol{b}' := T\boldsymbol{b}$. Assume $a_1 < b_1'$. We can change exactly the first component of $\boldsymbol{b}'$ within the interval spanned by b_1' and $-b_1'$ using a convex linear combination of $\boldsymbol{b}'$ and $R_1\boldsymbol{b}'$. Here R_1 denotes the isometry which changes the sign of the first component. We have $a_1 \geqq 0$ and a_1 is in the mentioned interval. Hence we can reach $\boldsymbol{b}''$ with $\boldsymbol{b}'' \geqq \boldsymbol{a}$ and $(\boldsymbol{b}'')_1 = (\boldsymbol{a})_1$. After at most n steps we reach $\boldsymbol{a}$ which proves Theorem 14 (see [18, 38, 53, 55, 59]). ∎

1.5. An example: Comparison of Gibbsian states

With $\boldsymbol{a} \in l^1$ we associate a probability vector, called the *Gibbsian state given by* $\boldsymbol{a}$, with the aid of the formula

$$\varrho_{\boldsymbol{a}} := \{\varrho^1, \ldots, \varrho^n\} \quad \text{with} \quad \varrho^i = (z_{\boldsymbol{a}})^{-1} \exp(+a_i) . \tag{23}$$

z_a is a number, usually called a *partition sum*, given by

$$z_a = \sum_{i=1}^{n} \exp(+a_i) . \tag{24}$$

In statistical physics, $-a_i$ is typically the "i^{th} energy level" multiplied by the inverse of $k_B T$, where k_B is Boltzmann's constant and T is the absolute temperature.

Theorem 1-16. *Given* $\boldsymbol{a}, \boldsymbol{b} \in \boldsymbol{ml}^1$. *If*

$$a_i - a_{i+1} \geqq b_i - b_{i+1} , \qquad i = 1, 2, \dots ,$$

we have

$$\varrho_b \succ \varrho_a .$$

Proof. We abbreviate

$$\alpha_k = \sum_{i=1}^{k} \exp(+a_i); \qquad \tilde{\alpha}_k = \sum_{i=1}^{k+1} \exp(+a_i)$$

so that

$$e_k(\varrho_a) = \alpha_k(\alpha_k + \tilde{\alpha}_k)^{-1} .$$

For Gibbsian states $e_n = 1$ by normalization. Hence, denoting the corresponding quantities of ϱ_b by the letter β, our assertion is equivalent to

$$\tilde{\alpha}_k/\alpha_k \leqq \tilde{\beta}_k/\beta_k , \qquad k = 1, 2, \dots \tag{*}$$

Setting $d_j = a_j - b_j$, we have $d_1 \geqq d_2 \geqq \cdots$. Now

$$\alpha_k = \sum_{i=1}^{k} (\exp d_i)(\exp b_i) \geqq (\exp d_k)\, \beta_k .$$

In the same way we get $\tilde{\alpha}_k \leqq (\exp d_{k+1})\, \tilde{\beta}_k \leqq (\exp d_k)\, \tilde{\beta}_k$. Hence (*) is valid (see [95]). ∎

Here we remark on two simple situations where the theorem applies to physics: The raising of the temperature in a Gibbsian state gives more and more chaotic Gibbsian states (see [95, 104]). The same is true with Gibbsian states describing an isotherm expansion of a free gas. (Strictly speaking one had to use a simple extension of Theorem 16 covering the case $n = \infty$.)

1.6. An example: Localization of spectra for Hermitian matrices

We consider Hermitian matrices $A, B, \dots, Q, \dots$ of dimension m and assume $m \leqq n = \dim \boldsymbol{l}^1$, sometimes even $2m \leqq n$.

We denote by

$$\text{Spec}\, A := \{a_1, a_2, \dots, a_m, 0, 0, \dots, 0\} \in \boldsymbol{l}^1$$

the eigenvalues of A in decreasing order, with correct multiplicity, and supplemented by zeros to gain an element of $\boldsymbol{l}^1$. As an example, $Q = Q^*$ is a one-dimensional projection iff Spec $A = \{1, 0, 0, \dots, 0\}$. Further, Spec $A \in \boldsymbol{ml}^1$ iff $A \geqq 0$.

Theorem 1-17. *Let* A *be a positive semidefinite matrix and consider an element* $\boldsymbol{b} = \{b_1, \dots, b_n\} \in \boldsymbol{l}^1$ *with* $\boldsymbol{b} \geqq 0$. *If there are one-dimensional projections* $Q_1, Q_2, \dots$ *with*

$$A = \sum_i b_i Q_i \tag{25}$$

then it follows

$$\boldsymbol{b} \succ \operatorname{Spec} A . \tag{26}$$

Proof. Denote by N the set of all $\boldsymbol{b} \geqq 0$ for which (25) can be fulfilled with suitable one-dimensional projections. $\boldsymbol{c} \in N$ is called *minimally chaotic in N* iff $\boldsymbol{b} \in N$ and $\boldsymbol{c} \succ \boldsymbol{b}$ implies $\boldsymbol{c} \sim \boldsymbol{b}$, i.e. implies the existence of a permutation map P with $\boldsymbol{b} = P\boldsymbol{c}$.

Now we assume with $\boldsymbol{b} \in N$ and one-dimensional projections

$$A = \sum b_k Q_k$$

and for the two indices i, j, the inequality $b_i \geqq b_j > 0$ and $Q_i Q_j \neq Q_j Q_i$. We want to show the impossibility for $\boldsymbol{b}$ to be minimally chaotic in N. Indeed, there are two one-dimensional projections Q' and Q'', and two numbers $b' \geqq b'' > 0$ with

$$b_i Q_i + b_j Q_j = b'Q' + b''Q'' \quad \text{and} \quad Q'Q'' = 0 , \tag{$*$}$$

for exactly two of the eigenvalues of the matrix on the left-hand side of $(*)$ are different from zero, and they coincide with b', b''. Performing the trace in $(*)$ we arrive at

(a) $\qquad b_i + b_j = b' + b''$.

Performing the trace in $(*)$ after squaring we get

(b) $\qquad (b_i)^2 + (b_j)^2 < (b')^2 + (b'')^2$

for our assumption implies $Q_i Q_k \neq 0$ which gives for positive semidefinite matrices Tr. $Q_i Q_k > 0$. Now (a) and (b), and the positivity of the numbers involved, give rise to

$$b' > b_i \geqq b_j > b'' > 0 . \tag{$**$}$$

We denote by $\boldsymbol{b}'$ the element of l^1 gained by substituting b_i by b' and b_j by b''. By construction we have $\boldsymbol{b}' \in N$. Furthermore, $(**)$ ensures the existence of an elementary doubly stochastic map T with $T\boldsymbol{b}' = \boldsymbol{b}$. Hence $\boldsymbol{b}' \prec \boldsymbol{b}$ and the construction makes certain that $\boldsymbol{b}$ is not a permutation of $\boldsymbol{b}'$. Consequently, every minimally chaotic in N element $\boldsymbol{c}$ gives rise to a representation of A with mutually one-to-another orthogonal one-dimensional projections, i.e. to a spectral decomposition of A. (The projections, multiplied with a zero component, $b_k = 0$, do not play any role.)

Working in finite-dimensional spaces only, it is not hard to see that N is compact. We leave as a simple exercise (use for instance Theorem 13) the fact that in a compact set N to every $\boldsymbol{b} \in N$ there is at least one minimally chaotic in N element $\boldsymbol{c}$ with $\boldsymbol{c} \prec \boldsymbol{b}$. As shown above, in our case, $\boldsymbol{c}$ necessarily belongs to a spectral decomposition of A (see [93]). ∎

As a matter of fact, (26) vice versa implies the existence of a representation (25). But we shall not bother to prove this uninteresting statement. Rather, we shall give some implications of the theorem.

To this end we first extend the notations (15), (18), to Hermitian matrices A (not necessarily definite ones). If $(\dim A) \geqq s \geqq 0$

$$e_s(A) = \sup_D \operatorname{Tr.} AD \tag{27}$$

where D runs through all the matrices

$$0 \leqq D \leqq 1\,, \qquad \text{Tr.}\, D = s\,, \tag{27a}$$

1 denoting the unity matrix. For larger s we define

$$e_s(A) = \text{Tr.}\, A \qquad \text{for} \qquad s \geqq (\dim A)\,. \tag{27b}$$

To become acquainted with this notation we note the analogue to Lemma 12 due to Ky Fan and Horn (see [30, 31, 39]).

Lemma 1-18. *For every Hermitian matrix A we have*

$$e_s(A) = e_s(\text{Spec}\, A)\,. \tag{28}$$

Proof. A acts on an appropriate Hilbert space $\boldsymbol{l}^2$, and in this space there is a complete orthonormal system $\xi_1, \xi_2, \ldots, \xi_m$ of eigenvectors of A: $A\xi_k = a_k\xi_k$. We get

$$e_s = \sup \sum a_k(\xi_k, D\xi_k)$$

where $(.,.)$ is the scalar product of our $\boldsymbol{l}^2$. Now $\forall j$: $t_j = (\xi_j, D\xi_j)$ satisfies (15a), and every n-tupel (15a) can be represented in this way using a suitable (even with A commuting) D fulfilling (27a). On the other hand, (27b) is correct by convention. ∎

For natural $k \leq (\dim A)$ the functional $e_k(A)$ equals the sum of the k largest eigenvalues of A.

It is worthwhile mentioning an easy consequence of definitions (15) and (27) which we shall use several times later.

Lemma 1-19. *Both functions*

$$s \to e_s(\boldsymbol{a}) \quad \textit{and} \quad s \to e_s(A)$$

are concave and monotonously increasing ones for $s \geqq 0$. In $\boldsymbol{a}$ (resp. in A) they are subadditive,

$$e_s(\boldsymbol{a} + \boldsymbol{b}) \leqq e_s(\boldsymbol{a}) + e_s(\boldsymbol{b})\,, \tag{29}$$

$$e_s(A + B) \leqq e_s(A) + e_s(B) \tag{29a}$$

and for real, non-negative numbers t

$$e_s(t\boldsymbol{a}) = te_s(\boldsymbol{a}) \quad \textit{and} \quad e_s(tA) = te_s(A)\,. \tag{30}$$

Namely, e_s is represented as a supremum of linear in $\boldsymbol{a}$ (resp. in A) functionals. This proves (29) and (30). The increasing in s is obvious. To see the concavity in s for the matrix case, we arbitrarily choose D_1 and D_2 with Tr. $D_j = s_j$, $0 \leq D_j \leqq 1$. For $0 \leqq t \leqq 1$ we define $s = ts_1 + (1 - t)\, s_2$ and $D = tD_1 + (1 - t)\, D_2$. Because of $0 \leqq D \leqq 1$ and Tr. $D = s$ we get

$$e_s(A) \geqq t\, \text{Tr.}\, D_1A + (1 - t)\, \text{Tr.}\, D_2A$$

and, using the arbitrariness in the choice of D_1 and D_2,

$$e_s(A) \geqq te_{s_1}(A) + (1 - t)\, e_{s_2}(A)\,.$$

The concavity of $s \to e_s(\boldsymbol{a})$ follows in the same manner. ∎

Remind that (29) and (30) implies convexity in $\boldsymbol{a}$ (resp. in A) of the functionals e_s. This and (29a) are due to Horn and Ky Fan. Remarkably enough, Theorem 17 provides us with inequalities going in the opposite direction to (29):

Theorem 1-20. *Let k, i, j be natural numbers with $k = i + j$, and denote by A, B two positive semidefinite matrices. The sum of the k largest eigenvalues of $A + B$ is not smaller than the sum of the i largest eigenvalues of A plus the sum of the j largest eigenvalues of B.*

More generally, for all $s \geqq 0$ and $t \geqq 0$

$$e_{s+t}(A + B) \geqq e_s(A) + e_t(B) . \tag{31}$$

Proof. We denote by $a_1, a_2, \dots$ and $b_1, b_2, \dots$ the eigenvalues of A and B respectively. Just adding the spectral representations of A and of B we see from Theorem 17

$$\text{Spec}\,(A + B) \prec \{a_1, \dots, a_m, b_1, \dots, b_m\} := \boldsymbol{c} . \tag{0}$$

Hence by (28) and Theorem 13

$$e_k(A + B) \geqq e_k(\boldsymbol{c}) .$$

But $e_k(\boldsymbol{c})$ is the sum of the k largest components of $\boldsymbol{c}$. Therefore, $e_k(\boldsymbol{c}) = \sup\,[e_i(\boldsymbol{a}) + e_j(\boldsymbol{b})]$ under the condition $k = i + j$. Now with the aid of (17) we conclude from this relation $e_{s+t}(\boldsymbol{c}) \geqq e_s(\text{Spec}\,A) + e_t(\text{Spec}\,B)$. Thus (31) follows on equally well from (0) for $k = s + t$ (comp. (22)). ▮

Note that (31) becomes trivial if $s + t$ is larger or equal to the number of nonzero eigenvalues of $A + B$.

With some more knowledge of the connection between convexity and the relation $\succ$ we shall see further consequences of Theorem 17 later on.

1.7. Convexity and the relation $\succ$

A subset N of l^1 is called *symmetric* iff $\boldsymbol{a} \in N$ implies $P\boldsymbol{a} \in N$ for every permutation map P. N is called *isometrically invariant* iff $\boldsymbol{a} \in N$ implies $T\boldsymbol{a} \in N$ for all isometries of l^1.

Accordingly, a functional $\boldsymbol{a} \to F(\boldsymbol{a})$, defined on a symmetric (resp. isometrically invariant) set N, is said to be *symmetric* (resp. *isometrically invariant*) iff $F(\boldsymbol{a}) = F(T\boldsymbol{a})$ for all $\boldsymbol{a} \in N$ and all permutation maps (resp. for all isometries) T.

More explicitly, symmetricity of $F(\boldsymbol{a}) = F(a_1, \dots, a_n)$ means

$$F(a_1, a_2, \dots, a_n) = F(a_{i_1}, a_{i_2}, \dots, a_{i_n}) \tag{32}$$

for every permutation $j \to i_j$ and all $\boldsymbol{a} = \{a_1, a_2, \dots, a_n\}$ in its domain of definition. If, in addition,

$$F(a_1, a_2, \dots, a_n) = F(|a_1|, |a_2|, \dots, |a_n|) \tag{33}$$

is valid, then F turns out to be isometrically invariant (Lemma 1).

Let K be a convex subset of l^1. A real-valued functional F defined on K is a *convex* one iff it is simultaneously convex in all the components of its argument $\boldsymbol{a} \in K$, i.e. iff for all $\boldsymbol{a}, \boldsymbol{b} \in K$ and for all $t \in \mathbf{R}$ with $0 \leqq t \leqq 1$:

$$F\big(t\boldsymbol{a} + (1 - t)\,\boldsymbol{b}\big) \leqq tF(\boldsymbol{a}) + (1 - t)\,F(\boldsymbol{b}) . \tag{34}$$

The same remark applies to the definition of *concave* functionals, i.e. functionals F for which $-F$ is convex or, what is the same, for which the inequality sign is reversed in (34).

The following theorem, if combined with Theorem 13 is in [30, 66].

Theorem 1-21. *Let K be a symmetric and convex subset of $\boldsymbol{l}^1$ and F a symmetric and convex functional defined on K. If $\boldsymbol{a} \in K$ and $\boldsymbol{b} \succ \boldsymbol{a}$, then $\boldsymbol{b} \in K$ and*

$$F(\boldsymbol{b}) \leqq F(\boldsymbol{a}) .$$

(The inequality sign is reversed if "convex functional" is replaced by "concave functional".)

Proof. Indeed, there is a doubly stochastic T with $\boldsymbol{b} = T\boldsymbol{a}$ by assumption, and by Theorem 13 we have $T = \sum t_i T_i$, i.e. T is represented by a convex sum of permutation maps $T_1, T_2, \ldots$ Hence, first using convexity and then symmetry, we not only get $\boldsymbol{b} \in K$ but also

$$F(\boldsymbol{b}) \leqq \sum t_j F(T_j \boldsymbol{a}) = \sum t_j F(\boldsymbol{a}) = F(\boldsymbol{a})$$

q.e.d. ▌

Clearly, we may use Theorem 14 to prove the following theorem in the same way.

Theorem 1-22. *Let K be a convex and isometrically invariant subset of $\boldsymbol{l}^1$, and assume the convex and isometrically invariant functional F to be defined on K. If $\boldsymbol{b} \in K$ and*

$$\boldsymbol{a} \geqq 0 \quad \textit{and} \quad e_j(\boldsymbol{a}) \leqq e_j(\boldsymbol{b}) \quad \textit{for} \quad j = 1, 2, \ldots, n$$

it is

$$\boldsymbol{a} \in K \quad \textit{and} \quad F(\boldsymbol{a}) \leqq F(\boldsymbol{b}) .$$

The *proof* is as for the preceding theorem and we only note the 'compensation' of the somewhat stronger requirements for F by the possibility of having $e_n(\boldsymbol{a}) \neq e_n(\boldsymbol{b})$. [48]

An important situation in which Theorem 22 applies, arises for such F which are isometrically invariant norms of $\boldsymbol{l}^1$. For such functionals the statement of the theorem has been proved by von Neumann [61], see also [48].

Remark. A symmetric functional which is monotonous with respect to $\succ$, is *not* necessarily a convex one. This we can see very easily. Assume $s \to f(s)$ to be monotonous and $\boldsymbol{a} \to F(\boldsymbol{a})$ to be symmetric and convex. Then $\boldsymbol{a} \to \tilde{F}(\boldsymbol{a}) := f(F(\boldsymbol{a}))$ needs not be convex nor concave. But from Theorem 21 we conclude from $\boldsymbol{b} \succ \boldsymbol{a}$ either $\tilde{F}(\boldsymbol{a}) \geqq \tilde{F}(\boldsymbol{b})$ for increasing f or $\tilde{F}(\boldsymbol{a}) \leqq \tilde{F}(\boldsymbol{b})$ for decreasing f. These simple considerations apply to the so-called α-entropies defined below by formula (46). ▌

We now construct an important class of convex (concave) and symmetric functionals. In doing so, we restrict ourselves to the domain of definition

$$\boldsymbol{l}^1_+ := \{\boldsymbol{a} \in \boldsymbol{l}^1 : \boldsymbol{a} \geqq 0\} \tag{35}$$

leaving aside more general domains of definition which can be handled, using trivial modifications only, in the same way.

Definition 1-3. Let $s \to f(s)$ be a real-valued function defined for all $s \geqq 0$ and satisfying $f(0) = 0$. We define

$$S_f(\boldsymbol{a}) := \sum_{j=1}^{n} f(a_j) \quad \text{for} \quad \boldsymbol{a} = \{a_1, \ldots, a_n\} \in \boldsymbol{l}^1_+ . \tag{36}$$

For every positive semidefinite matrix A we define

$$S_f(A) := S_f(\text{Spec } A) . \tag{37}$$

The condition $f(0) = 0$ in this definition is essential only in a few cases and its influence can, if necessary, easily be checked. Another remark is the possibility of rewriting (37) with the help of the usual definition of a function $A \to f(A)$ of a Hermitian matrix:

$$S_f(A) = \text{Tr. } f(A) . \tag{37a}$$

The functionals of type S_f turn out to be relevant for convex and concave f. Especially, for $f(s) = -s \ln s$ if $s > 0$ and $f(0) = 0$ we get, up to the Boltzmann factor, the *entropy functional*. For a good and readable discussion of this entropy of states see [89, 109] and also [32]. The following theorems show properties of the entropy functional which are shared by all S_f with concave f. For this reason, S_f with concave f is sometimes called an "entropy-like" functional. To get the formulation with convex functions in the following theorems we only have to remember that $-f$ is convex iff f is concave and the relation $S_{(-f)} = -S_f$.

Theorem 1-23. *For all* $\boldsymbol{a}, \boldsymbol{b} \in \boldsymbol{l}^1_+$ *and all in* $0 \leqq s$ *defined concave functions* $s \to f(s)$ *satisfying* $f(0) = 0$ *we have with* $0 \leqq t \leqq 1$

$$S_f\big(t\boldsymbol{a} + (1-t)\,\boldsymbol{b}\big) \geqq tS_f(\boldsymbol{a}) + (1-t)\,S_f(\boldsymbol{b}) \tag{38}$$

and

$$S_f(\boldsymbol{a} + \boldsymbol{b}) \leqq S_f(\boldsymbol{a}) + S_f(\boldsymbol{b}) . \tag{39}$$

There is, indeed, a sharper result: The assertions of the theorem even hold for positive semidefinite matrices. Choosing all matrices in the following theorem to be diagonal we arrive at Theorem 23 above. Thus we only need to prove:

Theorem 1-24. *For all positive semidefinite matrices* A, B, *and all in* $0 \geqq s$ *defined concave functions* $s \to f(s)$ *satisfying* $f(0) = 0$ *we have with* $0 \leqq t \leqq 1$

$$S_f\big(tA + (1-t)\,B\big) \geqq tS_f(A) + (1-t)\,S_f(B) \tag{40}$$

and

$$S_f(A + B) \leqq S_f(A) + S_f(B) . \tag{41}$$

The concavity (40) has been proved by J. von Neumann [60]. The following one is not the shortest proof. But on its way we recover a useful inequality (42) and a lemma which is slightly more general than inequality (40): We look at the Hermitian matrices as Hermitian operators acting on a suitable (finite-dimensional and complex) Hilbert space $\boldsymbol{l}^2$. If $\xi \in \boldsymbol{l}^2$ and if $\eta_1, \eta_2, \dots$ form a complete orthonormal system of eigenvectors of a positive semidefinite operator A, then we have by concavity of f

$$f\big((\xi, D\xi)\big) = f(\Sigma\, t_j a_j) \geqq \Sigma\, t_j f(a_j) .$$

Here

$$t_j := |(\xi, \eta_j)|^2 \quad \text{and} \quad A\eta_j = a_j \eta_j .$$

But $f(a_j)$ is the expectation value $\big(\eta_j, f(A)\,\eta_j\big)$ of $f(A)$ with respect of η_j and, substituting this into the right-hand side of the inequality above, we are left with

$$\forall \text{ concave } f\colon\ f\big((\xi, A\xi)\big) \geqq \big(\xi, f(A)\,\xi\big) . \tag{42}$$

Clearly, this inequality will be reversed for convex functions (see [48, 96, 105, 106]).

Let us use this for the convex function $g := -f$ and let us remember definition (27) of e_s for $s \leqq \dim A$. For all Hermitian matrices D with $0 \leqq D \leqq 1$ and $\mathrm{Tr.}\, D = s$ we get

$$e_s(g(A)) \geqq \mathrm{Tr.}\, g(A)\, D = \sum d_j (\xi_j, g(A)\, \xi_j)$$

where $\xi_1, \xi_2, \ldots$ denotes a complete set of eigenvectors with eigenvalues $d_1, d_2, \ldots$ of D. Now (42), applied to $f = -g$, gives us to

$$e_s(g(A)) \geqq \sum d_j g((\xi_j, A\xi_j)) \,. \qquad (*)$$

Here we are allowed to choose the complete orthonormal system $\xi_1, \xi_2, \ldots$ and the numbers $d_1, d_2, \ldots$ arbitrarily satisfying

$$0 \leqq d_j \leqq 1 \quad \text{and} \quad \sum d_i = s \,.$$

Taking the supremum over all these possible choices we arrive at $e_s(g(A))$: We have only to take a complete orthonormal system of eigenvectors of A for $\xi_1, \xi_2, \ldots$ to see that the supremum of the right-hand side of $(*)$ is not smaller than $e_s(\mathrm{Spec}\, g(A))$ according to (15). But, according to Lemma 18, we have $e_s(\mathrm{Spec}\, g(A)) = e_s(g(A))$. Now we take the following into account: For every allowed choice of $\xi_1, \xi_2, \ldots$ and $d_1, d_2, \ldots$ the right-hand side of $(*)$ is convex: $A \to (\xi_j, A\xi_j)$ is linear and thus $A \to g((\xi_j, A\xi_j))$ is convex. In this way we see that the left-hand side of $(*)$ is a supremum of convex functionals and therefore convex itself. Let us fix this result (see [96]):

Lemma 1-25. *Let $t \to g(t)$ be a convex function. Then on the set of all Hermitian matrices with eigenvalues in the domain of definition of g*

$$A \to e_s(g(A)) \,, \qquad s \geqq 0 \,,$$

is convex.

We remark $e_s = e_n$ if $s \geqq \dim A$ and

$$e_s(g(A)) = \mathrm{Tr.}\, g(A) \quad \text{with} \quad s = \dim A \,.$$

So we have shown the convexity of $A \to S_g(A)$ which implies the concavity of $A \to -S_g(A) = S_{-g}(A)$. This proves (40) because of $f = -g$.

Now we come to assertion (41). We use the convention $n = 2 \dim A$. We know from Theorem 20

$$\mathrm{Spec}\,(A + B) \prec \{a_1, \ldots, a_m, b_1, \ldots, b_m\} := \boldsymbol{c} \qquad (**)$$

if the numbers a_i and b_j denote the eigenvalues of A and of B respectively. We now use (38) which has already been proved: S_f is concave on $\boldsymbol{l}^1_+$ and it is symmetric by its very definition (36). This guarantees the applicability of Theorem 21:

$$S_f(\mathrm{Spec}\,(A + B)) \leqq S_f(\boldsymbol{c}) \,.$$

Now we are done: The left-hand side equals $S_f(A + B)$ by Lemma 18, and by the same lemma together with (36) we see the right-hand side to be equal $S_f(A) + S_f(B)$ (see [93]).

Let us now consider a special class of concave functions (see [73, 104]). For real t we define

$$\left.\begin{aligned} g_t(s) &= 0 \quad &\text{if} \quad s \leqq t\,, \\ g_t(s) &= t - s \quad &\text{if} \quad s \geqq t\,. \end{aligned}\right\} \tag{43a}$$

Clearly, $s \to g_t(s)$ is a concave function. We abbreviate

$$\hat{e}_t(\boldsymbol{a}) := S_{g_t}(\boldsymbol{a}) \quad \text{and} \quad \hat{e}_t(A) := \hat{e}_t(\operatorname{Spec} A) \tag{43b}$$

so that

$$\hat{e}_t(\boldsymbol{a}) = \sum_i g_t(a_i) = \sum_{t<a_j} (t - a_j)\,. \tag{43c}$$

We have introduced these functionals because they are connected with the functionals e_s by a transformation looking similar to Legendre's. Indeed, let us calculate

$$st - e_s(\boldsymbol{a}) = st - \sup \sum s_i a_i = \inf \sum s_i(t - a_i)$$

where the infimum runs over all n-tuples $s_1, \ldots, s_n$, with $0 \leqq s_i \leqq 1$ and $\sum s_i = s$. Hence

$$\inf_s \,[st - e_s(\boldsymbol{a})] = \inf \sum s_i(t - a_i)$$

where now the numbers $s_1, s_2, \ldots$ are only restricted by $0 \leqq s_i \leqq 1$. To get the infimum one obviously had to choose $s_i = 0$ if $t > a_i$ and $s_i = 1$ if $t < a_i$, i.e.

$$\forall t\colon \inf_{s \geqq 0} \,[st - e_s(\boldsymbol{a})] = \hat{e}_t(\boldsymbol{a})\,. \tag{44}$$

This we derived under the assumption $n \geqq s \geqq 0$. But $st - e_s$ is always increasing for $s \geqq n$. Hence the restriction $s \leqq n$ is not necessary and can be dropped.

Formula (44) is supplemented by the inverse one

$$\forall s \geqq 0\colon \inf_t \,[st - \hat{e}_t(\boldsymbol{a})] = e_s(\boldsymbol{a})\,. \tag{45}$$

To prove this we use (44) to see $st - e_t \geqq e_s$. It remains to show equality for suitable t. It is sufficient to show this for s between 0 and n. If $0 \leqq s \leqq n$ there is an integer j within $0 \leqq j \leqq s < j + 1 \leqq n$. We get (see proof of (44)) $st - e_t = e_s$ with $t = a_{j+1}$, and under the assumption $a_1 > a_2 \geqq \cdots \geqq a_n$, if $s_i = 1$ for $i \leqq j$ and if $s_i = 0$ for $i < j + 1$. We choose the remaining number s_{j+1} by $s_{j+1} = s - j$. Thus we have obtained the wanted equality and $s = \sum s_i$ according to the choice of the numbers s_i. ∎

For the purpose of the next theorem let us slightly extend Definition 1-3: Let J denote a closed interval of the real axis. If then $s \to f(s)$ is a concave (convex) function defined on J and if all the components of $\boldsymbol{a} \in l^1$ are contained in J we define $S_f(\boldsymbol{a}) := \sum f(a_j)$.

Now we are ready to prove

Theorem 1-26. *Let J be a closed interval of the real axis. Assume all the components of the elements $\boldsymbol{a}, \boldsymbol{b} \in l^1$ to be contained in J. Then the following three conditions are equivalent one to another:*

(i) $\boldsymbol{a} \succ \boldsymbol{b}\,.$

(ii) *For all concave functions $s \to f(s)$ defined on J*

$$S_f(\boldsymbol{a}) \geqq S_f(\boldsymbol{b}) .$$

(iii) $\forall t \in J: \quad \hat{e}_t(\boldsymbol{a}) \geqq \hat{e}_t(\boldsymbol{b})$

and, in addition,

$$\sum a_i = \sum b_i .$$

Proof. The set of all elements of $\boldsymbol{l}^1$ the components of which are contained in J is a convex symmetric set. Hence (ii) follows from (i) if $\boldsymbol{a} \to S_f(\boldsymbol{a})$ is concave on this set. This is clear, for every functional $\boldsymbol{a} \to f(a_j)$, with a_j the jth component of $\boldsymbol{a}$, is concave by the concavity assumption for f. Thus (i) implies (ii), and let us now assume the validity of (ii) to prove (iii). Firstly, $\hat{e}_t$ is a special functional of the form S_f with concave f. Secondly, applying (ii) with the functions f: $s \to s$ and f: $s \to -s$, we see $\sum a_i = \sum b_i$ as well. Now assume (iii). The proof of (45) tells us: It suffices to take the infimum over all $t \in J$ in this formula. Hence we can conclude $e_s(\boldsymbol{a}) \leqq e_s(\boldsymbol{b})$ for all $s \geqq 0$. Now we can rely on Theorem 1-13 to get (i). ∎

Some effort was needed to prove Theorem 24. The proof above shows this to be a reflection of non-commutativity: Its commutative version, Theorem 23, indeed, is almost trivial. In the next chapter, however, we shall see how to recover some important "non-commutative versions" out of the commutative ones.

We end our consideration of convexity and $\succ$ by pointing shortly at the so-called α-entropies $S^{(\alpha)}$. They have been examined by Rényi [68] ("information of order α"). The case $\alpha = 0$ is known as the Hartley entropy. For $\alpha = 2$ see also Fano [32]. The definition is

$$S^{(\alpha)}(\boldsymbol{a}) := (1 - \alpha)^{-1} \ln \sum (a_j)^\alpha \tag{46}$$

for all $\boldsymbol{a} \neq 0$ and $\boldsymbol{a} \geqq 0$ and for all non-negative numbers α different from 0, 1, and ∞. For the excluded values it is defined by the limit if α goes to these exceptional values. $S^{(0)}$ is the logarithm of the number of non-zero components of $\boldsymbol{a}$. Further $S^{(1)} = S_f$ with $f(s) = -s \ln s$, and $S^{(\infty)}(\boldsymbol{a}) = -\ln \max_j a_j$.

Now $s \to (s)^\alpha$ is concave for $0 < \alpha \leqq 1$ and convex for $1 \leqq \alpha$. Hence the remark after Theorem 22 applies.

Taking into account the monotonicity of the logarithm we immediately arrive at the following conclusion (see [56, 109]):

Lemma 1-25. *Let $\boldsymbol{b} \in \boldsymbol{l}^1_+$ and $\boldsymbol{b} \neq 0$. Then for all $\alpha \geqq 0$*

$$\boldsymbol{a} \succ \boldsymbol{b} \quad \textit{implies} \quad S^{(\alpha)}(\boldsymbol{a}) \geqq S^{(\alpha)}(\boldsymbol{b}) .$$

1.8. Examples: Some maps and processes

We know $\boldsymbol{b} \succ \boldsymbol{a}$, $\boldsymbol{a} \geqq 0$ implies $\boldsymbol{b} \geqq 0$ and $\|\boldsymbol{a}\|_1 = \|\boldsymbol{b}\|_1$. Hence we may restrict ourselves for most purposes on the symmetric convex set

$$\Omega := \{\boldsymbol{a} \in \boldsymbol{l}^1_+ : \|\boldsymbol{a}\|_1 = 1\}$$

which consists of all probability vectors of length n.

At first we consider some maps

$$T: \Omega \to \Omega .$$

Such a map is called *chaos-enhancing* (or "mixing-enhancing") iff $T\boldsymbol{a} \succ \boldsymbol{a}$ for all $\boldsymbol{a} \in \Omega$. In case of $T\boldsymbol{a} \prec \boldsymbol{a}$ we call T *chaos-reducing.*

Lemma 1-27. *Every affine chaos-enhancing map T is the restriction on Ω of a doubly stochastic map.*

It is well known that an affine map $T: \Omega \to \Omega$ is the restriction of a linear map T' of $\boldsymbol{l}^1$ ($\boldsymbol{l}^1$ is generated by $\boldsymbol{l}^1_+$ and every ray of this cone hits Ω in just one point). Because Ω is invariant for T', this map is positivity preserving and isometrically acting on $\boldsymbol{l}^1_+$. Now $\boldsymbol{e} = \{1/n, 1/n, ...\}$ is the only maximally chaotic element of Ω. Hence $T'\boldsymbol{e} = T\boldsymbol{e} = \boldsymbol{e}$. Now Theorem 8, case (b), applies. ▮

As an example consider the affine map

$$T\boldsymbol{a} := (n\boldsymbol{e} - \boldsymbol{a})\,(n-1)^{-1}\,, \qquad \boldsymbol{a} \in \Omega\,.$$

Obviously, Ω is invariant with respect of T and $\boldsymbol{e}$ is a fixed point of T, i.e. T is chaos-enhancing, T is the restriction on Ω of the doubly stochastic map T' given by

$$(T'\boldsymbol{a})_j = (a_1 + \cdots + a_{j-1} + a_{j+1} + \cdots + a_n)\,(n-1)^{-1}\,.$$

WEHRL has called attention to some non-linear maps (see [104, 109]).

Theorem 1-28. *Let f be a non-negative function defined on the unite interval and let us consider the following map T from Ω into Ω which we define componentwise by*

$$(T\boldsymbol{a})_j = f(a_j)\,\big(f(a_1) + \cdots + f(a_n)\big)^{-1}$$

if the denominator involved is different from zero and $T\boldsymbol{a} = \boldsymbol{a}$ if $\forall j$: $f(a_j) = 0$.

(i) *If $s \to f(s)/s$ is monotonously increasing, then T is chaos-reducing, i.e. $T\boldsymbol{a} \prec \boldsymbol{a}$ always.*

(ii) *If $s \to f(s)/s$ is monotonously decreasing and $s \to f(s)$ monotonously increasing, then T is chaos-enhancing, i.e. $T\boldsymbol{a} \succ \boldsymbol{a}$ always.*

Proof. Let us order the components of $\boldsymbol{a}$ decreasingly, $a_1 \geqq a_2 \geqq \cdots$ and let us abbreviate $b_j = f(a_j)$. Then in case (i) we have $a_j b_i \geqq a_i b_j$ for $i \leqq j$. We conclude

$$(b_1 + \cdots + b_k)\,(a_{k+1} + \cdots + a_n) \geqq (a_1 + \cdots + a_k)\,(b_{k+1} + \cdots + b_n)\,.$$

Adding on both sides $(b_1 + \cdots + b_k)\,(a_1 + \cdots + a_k)$ we arrive at

$$(b_1 + \cdots + b_k) \sum a_i = (a_1 + \cdots + a_k) \sum b_i\,. \qquad (*)$$

Reminding $\sum a_i = 1$ we get $e_k(\boldsymbol{b}) \geqq (b_1 + \cdots + b_k) \geqq e_k(\boldsymbol{a}) \sum b_j$ which is assertion (i). In case (ii) the sign in (*) is reversed. But here, due to the monotonicity of f, we can conclude $b_1 + \cdots + b_k = e_k(\boldsymbol{b})$. Hence also (ii) is true. ▮

Corollary. *Theorem* 28, (i), *applies for non-negative convex functions f satisfying* $f(0) = 0$. (ii) *applies for non-negative, concave and monotonously increasing f.*

Namely, under these conditions $f(s)/s$ is increasing (decreasing) for convex (concave) functions f. ▮

Interpreting Ω as the set of states of a physical system, a concrete evolution of this system is described by a directed path in Ω which is parametrized by a time-parameter. Hence let us call *process* a map of the real half axis $0 \leqq t$ into Ω

$$t \to \boldsymbol{a}_t \in \Omega\,, \qquad t \geqq 0\,.$$

If the states, i.e. the elements of Ω, become more and more chaotic in the course of time, i.e. iff

$$\boldsymbol{a}_t \succ \boldsymbol{a}_s \quad \text{whenever} \quad t \geqq s \geqq 0\,.$$

LASSNER and LASSNER [51] have called such a process "*c-process*". (There are also c-processes the parameter of which is not to be interpreted as time but, say, as temperature or volume. See the end of 1.5.)

In physics, important examples of c-processes are given by the solutions of a master equation: Let L be a linear map from $\boldsymbol{l}^1$ into $\boldsymbol{l}^1$ and let us consider the differential equation

$$L\boldsymbol{a}_t = (d/dt)\,\boldsymbol{a}_t\,. \tag{47}$$

Such an equation is called *master equation* if for every of its solution $t \to \boldsymbol{a}_t$ from $\boldsymbol{a}_0 \in \Omega$ it follows $\boldsymbol{a}_t \in \Omega$ for all $t \geqq 0$. Another way to put this is to require

$$t \to T_t := \exp tL\,, \qquad t \geqq 0\,, \tag{48}$$

to be a semi-group of stochastic mappings. It is well known and a simple exercise to show this to be true if and only if the matrix representation $L = (l_{ik})$ satisfies $\forall k$: $\sum_i l_{ik} = 0$ and $l_{ik} \geqq 0$ for $i \neq k$ (see [64]).

Now it is clear (Theorem 1.8): Under the condition $T_t\boldsymbol{e} = \boldsymbol{e}$, or, equivalently, $L\boldsymbol{e} = 0$, every solution $t \to \boldsymbol{a}_t$, $t \geqq 0$, with $\boldsymbol{a}_0 \in \Omega$ is a c-process (see [56, 57, 72, 73, 98, 99]).

Let us now consider a non-linear evolution equation, a carricature of Boltzmann's equation. This "discrete gas model" is described by an element $\boldsymbol{a} \in \Omega$, the jth component a_j of which is the probability of finding a particle in the jth phase space (or configuration space) cell. The transition probability, $A_{ijkl} \geqq 0$, gives the rate (per unit time) of transition of a pair of particles, one in the kth and the other in the lth cell, to be scattered with the resulting pair in the ith and j^{th} cell. Conservation of particles gives

$$\sum_{k,l} A_{ijkl} = \sum_{i,j} A_{ijkl} = 1\,, \tag{49}$$

i.e. double stochasticity with respect of index pairs.

Then, the equation reads

$$(d/dt)\,a_i = \sum_{j,k,l} (A_{ijkl}a_k a_l - A_{klij}a_i a_j)\,. \tag{50}$$

The sum $(a_1 + \cdots + a_n) = \text{const}$ is an integral of this equation, i.e. constant in time. Now we add a further assumption:

$$\sum_{j,k} A_{ijkl} = 1\,, \tag{51}$$

i.e. with probability one per unit time for given i, l, there are j, k that the transition $(i, j) \to (k, l)$ takes place. Then an idea of CRELL shows [113].

Lemma 1-29. *Let $t \to \boldsymbol{a}_t$ be a solution of* (50) *with subsidiary conditions* (49) *and* (51). *If $\boldsymbol{a}_0$ is an inner point of Ω then*

$$\forall t \geqq 0 : t \to \boldsymbol{a}_t$$

is a c-process.

Proof. We first apply (49) to simplify (50):

$$(d/dt)\, a_i = \Big(\sum_{j,k,l} A_{ijkl} a_k a_l\Big) - a_i\,.$$

Let now $s \to f(s)$ denote a convex function, defined on the whole real axis. Our aim is to show $(d/dt)\, S_f(\boldsymbol{a}) \leqq 0$. It is

$$(d/dt)\, S_f(\boldsymbol{a}) = \sum_i f'(a_i) \Big(\sum_{j,k,l} A_{ijkl} a_k a_l - a_i\Big)\,.$$

The convexity of f gives $(s - t)\, f'(t) \leqq f(s) - f(t)$. We use that inequality to get

$$(d/dt)\, S_f(\boldsymbol{a}) \leqq \sum_i \Big(f\Big(\sum_{j,k,l} A_{ijkl} a_k a_l\Big) - f(a_i)\Big)\,. \qquad (;)$$

We use twice the convexity of f in the following auxiliary inequality, at first using $\boldsymbol{a} \in \Omega$ and then (51):

$$\sum_i f\Big(\sum_{j,k,l} A_{ijkl} a_k a_l\Big) \leqq \sum_{i,l} a_l f\Big(\sum_{j,k} A_{ijkl} a_k\Big) \leqq \sum_{i,j,k,l} a_l A_{ijkl} f(a_k) = \sum_k f(a_k)$$

where the last equality can be obtained with the aid of (49) and of $\boldsymbol{a} \in \Omega$. Comparing this auxiliary inequality with (;), we arrive at the desired $(d/dt)\, S_f \leqq 0$ for $\boldsymbol{a} \in \Omega$.

Now we have only to mention $a_t \in \Omega$ for all $t \geqq 0$ if only $a_0 \in \Omega$. This is a result of CARLEMAN [114] and SIMONS [115]. Compare also [110].

1.9. A partial order of m-tuples $/a_1, \ldots, a_m/$

Our next task is motivated by two problems.

The first is connected with the *master equation* (47): We could apply our partial order $\succ$ only to those master equations having $\boldsymbol{e}$ as a stationary solution. In general, however, there will be a stationary solution different from $\boldsymbol{e}$. (For example, heat conducting in an anisotropic medium.) Can one 'deform' $\succ$ to a new partial order in such a way that the central role of $\boldsymbol{e}$ is now played by a given stationary solution? The second problem concerns our proof of the Rado Theorem via Lemma 1-15. This proof is a constructive one. There are, of course, much shorter existence proofs based on Hahn-Banach separation theorems. Being easier, this method of proving can be handled in much more complicated and complex situations. This will be seen below.

Let us return to a master equation (47)

$$L\boldsymbol{a}_t = (d/dt)\, \boldsymbol{a}_t$$

and let $\boldsymbol{a}_t, \boldsymbol{b}_t, \boldsymbol{c}_t, \ldots$ be some of its solutions. In case $t \leqq s$ there is a stochastic map $T = \exp\,(s - t)\, L$ with

$$T\boldsymbol{a}_t = \boldsymbol{a}_s\,, \qquad T\boldsymbol{b}_t = \boldsymbol{b}_s\,, \quad \ldots$$

If it happens that one of these solutions, say $\boldsymbol{c}_t$, is stationary and, even more, $\forall t$: $\boldsymbol{c}_t = \boldsymbol{e}$, then every solution of the master equation is a c-process inside Ω.

But what can be said in general cases? Coming from this question we introduce

Definition 1-7. Let $\boldsymbol{a}_1, \dots, \boldsymbol{a}_m, \boldsymbol{b}_1, \dots, \boldsymbol{b}_m$ denote $2m$ elements of $\boldsymbol{l}^1$. We write $/\boldsymbol{a}_1, \dots, \boldsymbol{a}_m/ \prec /\boldsymbol{b}_1, \dots, \boldsymbol{b}_m/$ if and only if there is a stochastic map, T, with

$$\boldsymbol{b}_j = T\boldsymbol{a}_j \quad \text{for} \quad j = 1, \dots, m .$$

Remarks. (1) The relation $/\boldsymbol{a}/ \succ /\boldsymbol{b}/$ has nothing to do with $\boldsymbol{a} \succ \boldsymbol{b}$. Indeed, the former is equivalent with $\sum a_i = \sum b_i$ and hence $/\boldsymbol{a}/ \sim /\boldsymbol{b}/$ for all $\boldsymbol{a}, \boldsymbol{b} \in \Omega$.

(2) It is $/\boldsymbol{b}, \boldsymbol{e}/ \succ /\boldsymbol{a}, \boldsymbol{e}/$ if and only if $\boldsymbol{b} \succ \boldsymbol{a}$. This follows from Theorem 1-8, for $T \in \mathrm{ST}$ and $T\boldsymbol{e} = \boldsymbol{e}$ iff T is doubly stochastic.

(3) If $\boldsymbol{d}$ is a stationary and $t \to \boldsymbol{a}_t$ an arbitrary solution of a master equation we certainly have $/\boldsymbol{a}_s, \boldsymbol{d}/ \succ /\boldsymbol{a}_t, \boldsymbol{d}/$ if $s \geqq t$. This was the reason for a notation $\boldsymbol{a} \overset{\boldsymbol{d}}{\succ} \boldsymbol{b}$ iff $/\boldsymbol{a}, \boldsymbol{d}/ \succ /\boldsymbol{b}, \boldsymbol{d}/$. Later it became evident how natural the more general Definition 1-4 is (eventually reduced to $m = 2$).

Our first aim is a statement, analogous to Theorem 1-21. We fix, as in Definition 1-4, a natural number m and we consider a m-dimensional *convex cone* $\boldsymbol{C}$, i.e. a subset of R^m with the following properties:

$$\forall t \geqq 0\colon (s_1, \dots, s_m) \in \boldsymbol{C} \quad \text{implies} \quad (ts_1, \dots, ts_m) \in \boldsymbol{C}, \tag{52a}$$

$$(s_1, \dots, s_m) \in \boldsymbol{C} \text{ and } (t_1, \dots, t_m) \in \boldsymbol{C} \text{ implies } (s_1 + t_1, \dots, s_m + t_m) \in \boldsymbol{C}. \tag{52b}$$

We now comment on a class of real valued functions

$$f\colon (s_1, \dots, s_m) \to f(s_1, \dots, s_m)$$

defined on a convex cone $\boldsymbol{C}$. The role of convex (concave) functions on general convex sets is for convex cones given to such convex (concave) functions, which are homogeneous of degree one in addition. Homogeneity and convexity gives subadditivity. Three important properties a function can have:

$$\forall t \geqq 0\colon f(ts_1, \dots, ts_m) = tf(s_1, \dots, s_m), \tag{53a}$$

$$f(s_1 + t_1, \dots, s_m + t_m) \leqq f(s_1, \dots, s_m) + f(t_1, \dots, t_m), \tag{53b}$$

$$f(s_1 + t_1, \dots, s_m + t_m) \geqq f(s_1, \dots, s_m) + f(t_1, \dots, t_m). \tag{53c}$$

For convenience we call a function, f, defined on a convex cone, *h-convex* (*h-concave*) iff f satisfies on its domain of definition (53a) and (53b) (respectively (53a) and (53c)). The sets of all these functions we occasionally denote by

h-convex $\boldsymbol{C}$ and h-concave $\boldsymbol{C}$

respectively.

The next step is in associating to every m-dimensional convex cone, $\boldsymbol{C}$, a subset $\boldsymbol{C}(n, m)$ of m-tuples $/\boldsymbol{a}_1, \dots, \boldsymbol{a}_m/$ of $\boldsymbol{l}^1 = \boldsymbol{l}^1_n$:

$$/\boldsymbol{a}_1, \dots, \boldsymbol{a}_m/ \in \boldsymbol{C}(n, m) \quad \text{iff for all } j, \quad 1 \leqq j \leqq n, \tag{54}$$

$$((\boldsymbol{a}_1)_j, (\boldsymbol{a}_2)_j, \dots, (\boldsymbol{a}_m)_j) \in \boldsymbol{C}$$

where $(\boldsymbol{a}_i)_j$ is the jth component of the element $\boldsymbol{a}_i \in \boldsymbol{l}^1$. In a natural way we can consider every m-tuple $/\boldsymbol{a}_1, \dots, \boldsymbol{a}_m/$ as an element of the direct sum

$$\boldsymbol{l}^1_{n,m} := \boldsymbol{l}^1_n \oplus \dots \oplus \boldsymbol{l}^1_n \quad (m \text{ summands}).$$

With respect to this linear space our set $\boldsymbol{C}(n, m)$ is a nm-dimensional convex cone.

Every linear map, T, from $\boldsymbol{l}^1$ into $\boldsymbol{l}^1$ gives rise to linear map, $T^{(m)}$, of $\boldsymbol{l}^1_{n,m}$ into $\boldsymbol{l}^1_{n,m}$ by the definition

$$T^{(m)}\colon\ /\boldsymbol{a}_1, \ldots, \boldsymbol{a}_m/ \to /T\boldsymbol{a}_1, \ldots, T\boldsymbol{a}_m/ .$$

With the aid of the matrix representation $T = (t_{jk})$ we write

$$\big((T\boldsymbol{a}_1)_j, \ldots, (T\boldsymbol{a}_m)_j\big) = \sum_k t_{jk}\big((\boldsymbol{a}_1)_k, \ldots, (\boldsymbol{a}_m)_k\big) . \tag{57}$$

It follows the stability of $\boldsymbol{C}(n, m)$ under every mapping $T^{(m)}$, provided T is positivity preserving:

$$T^{(m)}\colon \boldsymbol{C}(n, m) \to \boldsymbol{C}(n, m) \quad \text{for positivity preserving } T. \tag{58}$$

Starting from a convex cone $\boldsymbol{C}$ of dimension m we associate to every function, f, defined on $\boldsymbol{C}$ a new function, $S_f^{(m)}$, defined on $\boldsymbol{C}(n, m)$, quite similar as it is done by Definition 1-3.

Definition 1-5. *Let $(s_1, \ldots, s_m) \to f(s_1, \ldots, s_m)$ be a function defined on a m-dimensional convex cone $\boldsymbol{C}$. We define for $/\boldsymbol{a}_1, \ldots, \boldsymbol{a}_m/ \in \boldsymbol{C}(n, m)$*

$$S_f^{(m)}(\boldsymbol{a}_1, \ldots, \boldsymbol{a}_m) = \sum_j f\big((\boldsymbol{a}_1)_j, \ldots, (\boldsymbol{a}_m)_j\big) \tag{59}$$

where the sum is from $j = 1$ to $j = n$.

The definition provides us with h-convex (h-concave) $S_f^{(m)}$ for h-convex (h-concave) f, as is to be immediately seen from (59).

$$\left.\begin{aligned} f \in \text{h-convex } \boldsymbol{C} &\to S_f^{(m)} \in \text{h-convex } \boldsymbol{C}(n, m) ,\\ f \in \text{h-concave } \boldsymbol{C} &\to S_f^{(m)} \in \text{h-concave } \boldsymbol{C}(n, m) . \end{aligned}\right\} \tag{60}$$

Theorem 1-30. *Given the m-dimensional cone $\boldsymbol{C}$ and $/\boldsymbol{a}_1, \ldots, \boldsymbol{a}_m/ \in \boldsymbol{C}(n, m)$ the following conditions are equivalent:*

(i) $/\boldsymbol{b}_1, \ldots, \boldsymbol{b}_m/ \succ /\boldsymbol{a}_1, \ldots, \boldsymbol{a}_m/$,

(ii) $\forall f \in$ *h-convex* $\boldsymbol{C}$: $S_f^{(m)}(\boldsymbol{b}_1, \ldots, \boldsymbol{b}_m) \leqq S_f^{(m)}(\boldsymbol{a}_1, \ldots, \boldsymbol{a}_m)$,

(iii) $\forall f \in$ *h-concave* $\boldsymbol{C}$: $S_f^{(m)}(\boldsymbol{b}_1, \ldots, \boldsymbol{b}_m) \geqq S_f^{(m)}(\boldsymbol{a}_1, \ldots, \boldsymbol{a}_m)$.

Proof. Assume the validity of (i) and $\boldsymbol{b}_i = T\boldsymbol{a}_i$ with $T \in \mathrm{ST}$ für $1 \leqq i \leqq m$. Then by (57) and (59)

$$S_f^{(m)}(\boldsymbol{b}_1, \ldots, \boldsymbol{b}_m) = \sum_j f\Big(\sum_k t_{jk}(\boldsymbol{a}_1)_k, \ldots, \sum_k t_{jk}(\boldsymbol{a}_m)_k\Big) .$$

With $t_j = t_{j1} + \cdots + t_{jn}$ and h-convex f we may write

$$f\Big(\sum_k t_{jk}(\boldsymbol{a}_1)_k, \ldots\Big) = t_j f\Big(\sum_k t_j^{-1} t_{jk}(\boldsymbol{a}_1)_k, \ldots\Big) \leqq t_j \sum_k t_j^{-1} t_{jk} f\big((\boldsymbol{a}_1)_k, \ldots\big) .$$

Cancelling t_j and summing up the j-index (remember for this $t_{1k} + t_{2k} + \cdots + t_{nk} = 1$) gives the desired inequality (ii). (iii) follows in the same way from (i). It remains to disprove (i) if either (ii) or (iii) is wrong.

The set $N := \{T^{(m)}/\boldsymbol{a}_1, \ldots, \boldsymbol{a}_m/\colon T \text{ stochastic}\}$ is compact, convex, and by (58) a subset of $\boldsymbol{C}(n, m)$. If N does not contain $/\boldsymbol{b}_1, \ldots, \boldsymbol{b}_m/$ there is a linear functional, L, on $\boldsymbol{l}^1_{n,m}$ satisfying

$$L(\boldsymbol{b}_1, \ldots, \boldsymbol{b}_m) > \{\sup L(T\boldsymbol{a}_1, \ldots, T\boldsymbol{a}_m),\ T \in \mathrm{ST}\} . \tag{+}$$

(Replacing "sup" by "inf" and L by $-L$ we arrive at the opposite inequality which we use below without further comment.) The general form of L is well known: There exists m elements $\boldsymbol{t}_1, \dots, \boldsymbol{t}_m$ which represent L by

$$L(\boldsymbol{c}_1, \dots, \boldsymbol{c}_m) = \sum_{i,j} (\boldsymbol{c}_i)_j (\boldsymbol{t}_i)_j .$$

Let us now define

$$K^+(\boldsymbol{c}_1, \dots, \boldsymbol{c}_m; \boldsymbol{t}_1, \dots, \boldsymbol{t}_m) = \sup_T \sum_{i,j} (\boldsymbol{c}_i)_j (T\boldsymbol{t}_i)_j \tag{61}$$

where the supremum is over all stochastic T. The functional which would be defined by (61) if the operation "sup" is replaced by "inf" will be denoted by K^-.

From (+) we infer the validity of

$$K^+(\boldsymbol{b}_1, \dots, \boldsymbol{b}_m) > K^+(\boldsymbol{a}_1, \dots, \boldsymbol{a}_m) \quad \text{and} \quad K^-(\boldsymbol{b}_1, \dots) < K^-(\boldsymbol{a}_1, \dots) .$$

Hence we have reached more than we need by establishing the h-convexity of K^+ and the h-concavity of K^- on $\boldsymbol{l}^1_{n,m}$. As usual let $\boldsymbol{e}_k$ denote the element of $\boldsymbol{l}^1$ with $(\boldsymbol{e}_k)_k = 1$ and $(\boldsymbol{e}_k)_j = 0$ if $j \neq k$. Then $\boldsymbol{t}_i = \sum_k (\boldsymbol{t}_i)_k \boldsymbol{e}_k$. Now (61) can be written

$$\sup \sum_{i,j,k} (\boldsymbol{c}_i)_j (\boldsymbol{t}_i)_k (T\boldsymbol{e}_k)_j , \qquad T \in \mathrm{ST} .$$

To every choice of n elements $\boldsymbol{d}_1, \dots, \boldsymbol{d}_n$ of Ω there is exactly one stochastic T with $T\boldsymbol{e}_k = \boldsymbol{d}_k$, $k = 1, \dots, n$. If $\boldsymbol{d}$ is varying freely within Ω then (with k fixed)

$$\sup_{\boldsymbol{d}} \sum_j \left[\sum_i (\boldsymbol{c}_i)_j (\boldsymbol{t}_i)_k\right] (\boldsymbol{d})_j = \sup_j \sum_i (\boldsymbol{c}_i)_j (\boldsymbol{t}_i)_k .$$

Hence, (61) is converted into

$$K^+ = \sum_k \sup_j \sum_i (\boldsymbol{c}_i)_j (\boldsymbol{t}_i)_k .$$

Let us now call a function, f, defined on the whole m-dimensional real linear space an *elementary h-convex* one if there are m elements $\boldsymbol{t}_1, \dots, \boldsymbol{t}_m$ of $\boldsymbol{l}^1$ with

$$f(s_1, \dots, s_m) = \sup_k \sum_i s_i (\boldsymbol{t}_i)_k . \tag{62}$$

If in the definition (62) the operation "sup" is replaced by "inf" we call f *elementary h-concave*.

The result is

$$K^+ = S_f^{(m)} \quad \text{with } f \text{ defined by (62).} \tag{63}$$

This disproves (i) under the assumption that (ii) is wrong. In the same way, using K^- and an appropriate elementary h-concave function, the same assertion follows with respect to (iii). ∎

Corollary. *In order that*

$$S_f^{(m)}(\boldsymbol{b}_1, \dots, \boldsymbol{b}_m) \leq S_f^{(m)}(\boldsymbol{a}_1, \dots, \boldsymbol{a}_m)$$

is valid for every h-convex function f it is sufficient that this inequality is true for all elementary h-convex functions f.

Similarly, (iii) *of Theorem* 1-30 *is valid if this inequality is true for all elementary h-concave functions, f.*

An example: $m = 2$ (see [24, 56, 57, 74, 81, 99, 100, 101, 102]). Theorem 1-30 restricted to $m = 2$ and applied to the solutions of a master equation gives to us the so-called *H-theorems* of VAN KAMPEN and FELDERHOF. Namely, as already mentioned in the motivation at this paragraph's beginning: Any two solution a_t, b_t of a master equation satisfy

$$/a_t, b_t/ \succ /a_s, b_s/ \quad \text{if only} \quad t \geqq s \,.$$

Therefore, inserting two solutions in a h-concave f the resulting function of t is monotonously increasing with t.

For $m = 2$ one can simplify considerably the relevant functionals. Indeed, let $\boldsymbol{C}$ be the cone of all pairs (s_1, s_2) of real numbers satisfying $s_j > 0$. Every function, f, fulfilling (53a) may be written as follows

$$f(s_1, s_2) = s_2 g(s_1/s_2) \quad \text{with} \quad g(s) := f(s, 1) \,. \tag{64}$$

It is elementary to show that f in (64) is h-concave (or h-convex) iff g is a concave (resp. convex) function of the positive reals. Hence we get all relevant $S_f^{(2)}$ by

$$S_f^{(2)}(\boldsymbol{a}_1, \boldsymbol{a}_2) = \sum_j (\boldsymbol{a}_2)_j \, g\big((\boldsymbol{a}_1)_j/(\boldsymbol{a}_2)_j\big) \,. \tag{65}$$

Functionals of this form have been used by VAN KAMPEN [44, 45], and CSISZÁR [25]. VAN KAMPEN probably was the first who noticed its intime connection with master equations. The particular case $g := -s \ln s$ is well known as "relative entropy" (UMEGAKI [103]), "generalized Boltzmann-Gibbs-Shannon entropy" (SEGAL [83]), "information content" (KULLBACH [49]) and "relative information" (RÉNYI [68]). More recent discussions are found in [116, 117].

As in (65) we assume in the following all pairs contained in $\boldsymbol{C}(n, 2)$ with $\boldsymbol{C}$ as above.

In our special case $m = 2$ we can mimic the transformations (44) and (45) to get another characterization of $/\boldsymbol{a}_1, \boldsymbol{a}_2/ \succ /\boldsymbol{b}_1, \boldsymbol{b}_2/$. This we sketch now. Define as in (43a) g_t to be equal 0 if $s \leqq t$ and to be equal $t - s$ if $s \geqq t$. Define

$$\hat{e}_t(\boldsymbol{a}_1, \boldsymbol{a}_2) := S_f^{(2)}(\boldsymbol{a}_1, \boldsymbol{a}_2) \quad \text{with} \quad f = s_2 g_t(s_1/s_2) \,.$$

Using the fact that every concave g can be uniformly approximated on compact sets by positive linear combinations of the functions g_t and the general linear function $s \to \beta_1 s + \beta_2$, we see

$$/\boldsymbol{a}_1, \boldsymbol{a}_2/ \succ /\boldsymbol{b}_1, \boldsymbol{b}_2/ \quad \text{iff} \quad \forall t\colon \hat{e}_t^{(2)}(\boldsymbol{a}_1, \boldsymbol{a}_2) \geqq \hat{e}_t^{(2)}(\boldsymbol{b}_1, \boldsymbol{b}_2) \,. \tag{66}$$

Applying a Legendre-Young transformation we define for $s \geqq 0$

$$e_s^{(2)}(\boldsymbol{a}_1, \boldsymbol{a}_2) := \inf_t \, [st - \hat{e}_t^{(2)}(\boldsymbol{a}_1, \boldsymbol{a}_2)] \,. \tag{67}$$

It turns out that

$$\hat{e}_t^{(2)}(\boldsymbol{a}_1, \boldsymbol{a}_2) = \inf_{s \geqq 0} \, [st - e_s^{(2)}(\boldsymbol{a}_1, \boldsymbol{a}_2)] \,. \tag{68}$$

Because of this equivalence one can use the set of all functionals $\{e_s^{(2)}, s > 0\}$ to characterize $\succ$ for pairs:

$$/\boldsymbol{a}_1, \boldsymbol{a}_2/ \succ /\boldsymbol{b}_1, \boldsymbol{b}_2/ \quad \text{iff} \quad \forall s \geqq 0: e_t^{(2)}(\boldsymbol{a}_1, \boldsymbol{a}_2) \leqq e_s^{(2)}(\boldsymbol{b}_1, \boldsymbol{b}_2)\,. \tag{69}$$

One proves (68) either by a general argument or by similar elementary arguments as we used to prove (44) and (45). By the latter we prove even more, namely

$$e_s^{(2)}(\boldsymbol{a}, \boldsymbol{b}) = \sup \sum_j (\boldsymbol{a})_j\, t_j \tag{70}$$

where $\{t_1, \ldots, t_n\}$ runs through all sequences of real numbers satisfying

$$\forall i: 0 \leqq t_i \leqq 1 \quad \text{und} \quad \sum_j (\boldsymbol{b})_j\, t_j \leqq s\,. \tag{70a}$$

A more detailed description of the proof is in [101]. See also [47, 71, 77, 78].

2. Order structures of matrices

2.1. The relation $\succ$ in the space $\boldsymbol{B}^1$

By $\boldsymbol{B}^1$ or $\boldsymbol{B}^1_n$ we denote the linear space of all n-by-n-matrices equipped with the norm

$$A \to \|A\|_1 := \mathrm{Tr.}(A^*A)^{1/2} . \tag{1}$$

Here, as usual, $(A^*A)^{1/2}$ is the unique non-negative root of A^*A. For a proof of being a norm see [48] (or consult Theorem 2-6 below).

We write E for the unity matrix and denote by $\boldsymbol{B}^1_h$ the subspace of Hermitian matrices and by $\boldsymbol{B}^1_+$ the cone of positive semidefinite matrices. We have $\|A\|_1 = \mathrm{Tr.}\, A$ for all $A \in \boldsymbol{B}^1_+$.

A linear map

$$\boldsymbol{T}\colon \boldsymbol{B}^1 \to \boldsymbol{B}^1 \tag{2}$$

is called *positivity preserving* iff $\boldsymbol{T}\boldsymbol{B}^1_+ \subseteqq \boldsymbol{B}^1_+$.

Definition 2-1. A map $\boldsymbol{T}\colon \boldsymbol{B}^1 \to \boldsymbol{B}^1$ is called *stochastic* if it is linear, positivity preserving and trace preserving. The map $\boldsymbol{T}$ is called *doubly stochastic* iff it is a stochastic map satisfying $\boldsymbol{T}E = E$.

Definition 2-2. For two Hermitian matrices A, B, we write

$$A \succ B$$

and call A *more chaotic* (more mixed, less pure) *than* B iff there is a doubly stochastic map $\boldsymbol{T}\colon \boldsymbol{B}^1 \to \boldsymbol{B}^1$ with

$$A = \boldsymbol{T}B .$$

We write $A \sim B$ if and only if $A \succ B$ *and* $B \succ A$ (see [94]).

At once we see from this definition

$$A \succ B \quad \text{and} \quad B \succ C \quad \text{imply} \quad A \succ C . \tag{3}$$

To proceed further we need some notations. The Hermitian conjugate of a matrix A is denoted by A^*. The adjoint $\boldsymbol{T}^*$ of a linear map (2) is defined by

$$\forall A, B \in \boldsymbol{B}^1\colon \mathrm{Tr.}\, (\boldsymbol{T}A)\, B^* = \mathrm{Tr.}\, A(\boldsymbol{T}^*B)^* . \tag{4}$$

The existence and uniqueness of $\boldsymbol{T}^*$ may be seen as follows. $\boldsymbol{B}^1$, equipped with the scalar product

$$A, B \to \mathrm{Tr.}\, AB^*$$

becomes a finite-dimensional Hilbert space and T^* is the adjoint of T relative to this Hilbertian structure.

Next we need:

(i) If T is positivity preserving then so is T^*.

(ii) If T is trace preserving then $T^*E = E$.

(iii) If $TE = E$ then T^* is trace preserving.

so that altogether

(iv) If T is doubly stochastic, the same is true with T^*.

Indeed, if $B \geqq 0$ then we have by (4) and by positivity preserving Tr. $A(T^*B)^* \geqq 0$ for all $A \geqq 0$ which proves (i). Choosing $B = E$ in (4) and using trace preserving we find for all A the equation Tr. A = Tr. $A(T^*E)^*$ which is possible only if $T^*E = E$. Similarly one proves (iii). (iv) collects only (i), (ii), and (iii). $TE = E$ and positivity preserving shows further

(v) If $0 \leqq A \leqq E$ and T is doubly stochastic, then it follows $0 \leqq TA \leqq E$.

Lemma 2-1. *$A \sim B$ for two Hermitian matrices if and only if they are unitarily equivalent, i.e. iff there is a unitary matrix U with $A = UBU^{-1}$.*

The proof requires to show that Spec A = Spec B. This is a consequence of the assertion

$$A \prec B \quad \text{implies} \quad \text{Spec}\, A \prec \text{Spec}\, B\,. \tag{5}$$

Having in mind the trace preserving property of stochastic maps, (5) demands to show (see Lemma 1-18)

$$A \prec B \quad \text{implies} \quad \forall s \geqq 0\colon e_s(A) \leqq e_s(B)\,. \tag{6}$$

Assume now $A = TB$ with doubly stochastic T and denote by D_s the set of all matrices D satisfying $0 \leqq D \leqq E$ and Tr. $E = s$. (iv), trace preserving and (v) shows $T^*D_s \subseteqq D_s$. Hence, according to (1.27)

$$e_s(B) \geqq \text{Tr. } B(T^*D) = \text{Tr. } (TB)\,D = \text{Tr. } AD\,.$$

Taking the supremum over D_s we arrive at $e_s(A) \leqq e_s(B)$. It is a trivial fact that unitary equivalence of A and B implies $A \sim B$. (See also [16].) ∎

Theorem 2-2. *The following properties of a pair of Hermitian matrices, A, B, are mutually equivalent one to another:*

(a) $A \succ B$.

(b) *There are matrices $B_1, \ldots, B_m$, unitarily equivalent to B,*

$$\forall i\colon B_i \sim B \tag{7}$$

and non-negative real numbers $s_1, \ldots, s_m$ satisfying

$$\forall i\colon s_i \geqq 0 \quad \text{and} \quad \sum s_j = 1 \tag{8}$$

such that

$$A = \sum s_j B_j\,. \tag{9}$$

(c) *The same as in* (b) *but with the additional requirement*

$$\forall i, k: B_i B_k = B_k B_i . \tag{10}$$

(d) Spec $A \succ$ Spec B.

(e) Tr. $A =$ Tr. B *and* $e_j(A) \leqq e_j(B)$, $\quad j = 1, \ldots, n-1$.

(f) *For all concave functions* $s \to f(s)$ *defined on the real axis it is*

$$S_f(A) \geqq S_f(B) .$$

(g) *It is*

$$\text{Tr. } A = \text{Tr. } B \quad \textit{and} \quad \forall t: \hat{e}_t(A) \geqq \hat{e}_t(B)$$

where

$$\hat{e}_t(C) := \hat{e}_t(\text{Spec } C) .$$

Proof. According to Definition 1.3 we have $S_f(A) = S_f(\text{Spec } A)$. Hence by Theorem 1-26 and Theorem 1-13 the properties (d), (e), (f), and (g) are equivalent one to another. Further (c) $\Rightarrow$ (b), (b) $\Rightarrow$ (a) for the map

$$\boldsymbol{T}: C \to \sum s_j U_j C U_j^{-1}$$

where the unitaries U_i fulfill $B_i = U_i B U_i^{-1}$, is doubly stochastic. We have shown above the implication (a) $\Rightarrow$ (d). Hence it remains to see (d) $\Rightarrow$ (c). With a suitable unitary matrix $B' := UBU^{-1}$ is a diagonal matrix. Considering Spec A and Spec B to be elements of l^1, property (d) ensures us of the existence of numbers (8) and of permutation maps $P_1, P_2, \ldots$ with

$$\text{Spec } A = \sum s_j P_j \text{ Spec } B .$$

Clearly, Spec $B =$ Spec B' and every permutation P_i of the diagonal of B' (which can be identified with Spec B) can be implemented by a unitary transformation $B' \to U_i B' U_i^{-1} = B_i'$. The matrices B_i' are diagonal again, therefore commute, and

$$\text{Spec } A = \text{Spec } [\sum s_j B_j'] .$$

There is a unitary matrix V, therefore, with

$$V^{-1} A V = \sum s_j B_j'$$

and with $B_i := V B_i' V^{-1}$ and $s_1, s_2, \ldots$ as above all requirements for (c) have been fulfilled. (See [94, 95, 96].) ∎

Remark. In [94] the definition of $\succ$ has been given with the aid of property (b) of Theorem 2.2. In the next chapter we shall adopt a version of this property (b) to define $\succ$ under rather general circumstances.

Let us now write down some examples which are simple extensions of results obtained in Chapter 1.

Example 1. Let $A, B \in \boldsymbol{B}_h^1$ and denote by

$$a_1 \geqq a_2 \geqq \cdots \quad \text{and} \quad b_1 \geqq b_2 \geqq \cdots$$

their eigenvalues respectively. Then from

$$\forall i: a_i - a_{i+1} \geqq b_i - b_{i+1}$$

it follows that

$$(\exp A)/\mathrm{Tr.}\exp A \prec (\exp B)/\mathrm{Tr.}\exp B\,.$$

This is a mere rewriting of Theorem 1-16 with the help of Theorem 2.2.(d).

The matrix $(\exp C)/\mathrm{Tr.}\exp C$ with $C \in \boldsymbol{B}_h^1$ is called *Gibbsian density matrix given by C*. A *density matrix* is a matrix out of $\boldsymbol{B}_+^1$ with trace one. The set of all density matrices

$$\Omega^q := \{A \in \boldsymbol{B}_+^1 : \mathrm{Tr.}\,A = 1\} \tag{11}$$

is convex and compact. As is with Ω as defined in 1.8. the set Ω^q may be interpreted as the set of states of a (quantum) physical system. The index "q" in Ω^q should remind us of its occurrence in quantum statistics.

Example 2. Every linear map $\boldsymbol{L}: \boldsymbol{B}^1 \to \boldsymbol{B}^1$ gives rise to a differential equation

$$(d/dt)\,A_t = \boldsymbol{L}A_t\,. \tag{12}$$

Such an equation is called a (quantum) *master equation* if for every of its solution $t \to A_t$ from $A_0 \in \Omega^q$ it follows that $\forall t \geqq 0$: $A_t \in \Omega^q$. If in addition $\boldsymbol{L}E = 0$, then every solution of (12) with initial value $A_0 \in \Omega^q$ obeys

$$A_t \succ A_s \quad \text{if} \quad t \geqq s \geqq 0\,.$$

Namely, for $t \geqq 0$ the mapping $\exp t\boldsymbol{L}$ maps Ω^q into Ω^q if (12) is a master equation. If, in addition, $\boldsymbol{L}E = 0$ we have $(\exp t\boldsymbol{L})\,E = E$ and for non-negative t the map $\exp t\boldsymbol{L}$ is doubly stochastic.

It is sometimes convenient to convert the definition of a *process* to the case at hand to be a map of the positive half axis into Ω^q. In the same line of thinking we shall call such an object $t \to A_t$ a *c-process* iff $A_t \succ A_s$ for $t \geqq s \geqq 0$ is true.

Example 3. Using again criterion (d) of Theorem 2.2 we see the following from Theorem 1-28 (Wehrl). Let $s \to f(s)$ be defined in the unit interval and be non-negative there. Define

$$TA := A \qquad \text{if } \mathrm{Tr.}\,f(A) = 0\,,$$
$$TA := f(A)/\mathrm{Tr.}\,f(A) \qquad \text{otherwise}$$

for all $A \in \Omega^q$. Then

$A \succ TA$ if $s \to f(s)/s$ is monotonously increasing,

$TA \succ A$ if $s \to f(s)/s$ is monotonously decreasing and, in addition, $s \to f(s)$ is increasing.

2.2. Some doubly stochastic maps

Unfortunately the structure of a general positivity preserving map from $\boldsymbol{B}^1$ into $\boldsymbol{B}^1$ is not explicitly known. An important class of doubly stochastic mappings can be constructed in this way: Let $W_1, \dots, W_m$ be some matrices and define

$$TA := \sum W_j A W_j^*\,, \qquad A \in \boldsymbol{B}^1\,. \tag{13}$$

This is a positivity preserving map.

Remark. A positivity preserving map T is of the form (13) if and only if it is completely positive (KRAUS). The concept of complete positivity was introduced by STINESPRING and UMEGAGI and is, in many respects, more natural than mere positivity. However, we do not need the machinery associated with the concept of complete positivity. See [85, 86] for an introduction.

In order to be stochastic, a map of the form (13) has to preserve trace. This requirement reads

$$\sum W_j^* W_j = E . \tag{14a}$$

To be doubly stochastic, we further need $TE = E$, i.e.

$$\sum W_j W_j^* = E . \tag{14b}$$

Thus, (13) together with (14) and (15) gives us many examples of doubly stochastic maps. We have already come across such maps in the proof of Theorem 2.2:

$$A \to \sum s_j U_j A U_j^{-1}$$

with unitary U_i and probability vector $s_1, s_2, \ldots$, is after a trivial rewriting of form (13). A slightly more involved expression for a doubly stochastic map is occasionally of importance.

Let us consider a closed and hence (by finite-dimensionality) compact group G of unitary matrices. Denote by $d_G U$ the Haar measure of G which is unique after normalizing the Haar volume of G to one. Then

$$A \to A^G := \int_G U A U^{-1} \, d_G U \tag{15}$$

maps $\boldsymbol{B}^1$ linearly into itself. (In this definition we may understand the matrix elements $(A^G)_{ij}$ of A^G most simply as given by the Haar integral of $(UAU^{-1})_{ij}$.)

Without difficulty one can check positivity preserving, trace preserving and $E^G = E$, so that (15) defines a doubly stochastic map. More generally, we may replace $d_G U$ in the integral (15) by any other probability measure of G to get further examples of doubly stochastic maps.

The invariance properties of the Haar measure give rise to

$$A^G = U A^G U^{-1} = (UAU^{-1})^G \quad \text{for all } U \in G . \tag{16}$$

Lemma 2-3. *Let A be an arbitrary matrix and define*

$$K := \text{convex hull } \{UAU^{-1}, U \in G\} .$$

(i) *$B \in K$ implies $B^G \in K$.*

(ii) *$B \in K$ and $BU = UB$ for all $U \in G$ if and only if $B = A^G$.*

Proof. B^G is a limit of convex linear combinations

$$(1/m)\,(U_1 B U_1^{-1} + \cdots + U_m B U_m^{-1}) \in K , \qquad U_i \in G .$$

Hence assertion (i) is true if K is compact. But K is generated convexly by the compact set UAU^{-1}, $U \in G$. This gives, in finite-dimensional spaces, compactness of K (theorem of BOURBAKI). The "if-part" of (ii) is contained in (16). On the other hand if B commutes with all $U \in G$ we have $B = B^G$. Now B is a convex linear combination of matrices UAU^{-1}, $U \in G$ and by (16) each of these matrices are mapped by (15) onto A^G. Hence

$$B^G = A^G \quad \text{for all } B \in K \tag{17}$$

and (ii) is proved. (See [96].) ▮

Let us further note two properties of (15) which one gets straightforwardly:

$$(A^G)^* = (A^*)^G \quad \text{and} \quad \text{Tr. } A^G C = \text{Tr. } A^G C^G \tag{18}$$

for all matrices A and C.

Corollary 1. *Let $VA^GV^{-1} \in K$ with unitary V. Then*

$$VA^GV^{-1} = A^G .$$

Proof of the corollary. With $B = VA^GV^{-1}$ we certainly have $\text{Tr. } BB^* = \text{Tr. } A^G(A^G)^*$. Hence by (17) and (18)

$$\text{Tr. } (A^G - B)(A^G - B)^* = 2\,\text{Tr. } A^G(A^G)^* - \text{Tr. } A^G B^* - \text{Tr. } (A^G)^* B = 0 .$$

Corollary 2. *Let A be Hermitian. Then*

$$B \in K \quad \textit{and} \quad B \succ B' \quad \textit{for all} \quad B' \in K$$

if and only if $B = A^G$.

Proof of Corollary 2. (17) shows $A^G = B^G \succ B$ by double stochasticity of the map (15). Hence $B \succ A^G$ implies $B \sim A^G$, i.e. unitary equivalence (Lemma 2-1). Corollary 1 asserts $B = A^G$ in this situation. ▮

Lemma 2-4. *Let G be the smallest compact group which contains the two compact groups, G_1, G_2, of unitary matrices. For every $A \in \boldsymbol{B}^1$ the sequence ($i = 0, 1, 2, \ldots$)*

$$B_0 = A , \qquad B_{2i+1} = (B_{2i})^{G_1} , \qquad B_{2i+2} = (B_{2i+1})^{G_2}$$

is convergent and

$$\lim B_j = A^G .$$

Proof. By linearity of (15) we may restrict ourselves to Hermitian $A = A^*$. Then $B_{j+1} \succ B_j$ for all j. Hence $j \to e_s(B_j)$ is monotonously decreasing in j by Theorem 2-2. On the other side we have B_j contained in the convex hull K of all UAU^{-1} with $U \in G$, and K is compact (Lemma 2-3). The set $\boldsymbol{N}$ of limit points of the sequence $B_1, B_2, \ldots$ is not empty, therefore. For every $C \in \boldsymbol{N}$ we conclude $e_s(C) = \lim e_s(B_j)$. Together with the obvious $\text{Tr. } A = Tr.\ B_j = \text{Tr. } C$ we see: All matrices in $\boldsymbol{N}$ are unitarily equivalent. Now $C \in \boldsymbol{N}$ implies $C^{G_1} \in \boldsymbol{N}$. Our statement above gives $C \sim C^{G_1}$. Corollary 1 of Lemma 2-3 tells us now $C = C^{G_1}$. The same can be done with the group G_2. Hence C commutes with all matrices out of G_1 and G_2. This implies $UC = CU$ for all $U \in G$, and the application of Lemma 2-3 and (17) ensure us of $C = C^G = A^G$, i.e. $\boldsymbol{N}$ consists of A^G only. (See [96].) ▮

Corollary. *If under the same assumptions*

$$\forall U_1 \in G_1, \forall U_2 \in G_2{:} \quad U_1U_2 = U_2U_1$$

then

$$A^G = (A^{G_1})^{G_2} = (A^{G_2})^{G_1} .$$

Indeed, in that case, $B_j = B_2$ for all $j \geqq 2$ in the sequence mentioned in Lemma 2-4. ▮

One can give a rather explicit construction of each of the maps (15). We shall do this in several steps.

If G is given, we denote by $\check{G}$ the set of all unitary matrices depending linearly on G, i.e. $U \in \check{G}$ iff $U = \sum \lambda_j U_j$ with some $U_i \in G$ and complex numbers $\lambda_1, \lambda_2, \ldots$

$\check{G}$ is again a compact matrix group, which may be referred to as the *full group associated with* G. Especially, G is said to be *full* iff $G = \check{G}$.

A^G commutes with every element of $\check{G}$ and is certainly contained in the convex hull spanned by $\check{G}$, for $G \subseteqq \check{G}$ trivially. Thus Lemma 2.3 provides us with

$$A^G = A^{\check{G}} \quad \text{for all } A\,, \tag{19}$$

and it is sufficient to classify (15) for full groups G. ▮

Let us first assume G to be *commutative*. Then G is parametrized by, say, m reals, $s_1, \ldots, s_m$ and m orthoprojections $Q_1, \ldots, Q_m$,

$$G = \{U = \exp\,(s_1 Q_1 + \cdots + s_m Q_m) \quad \text{with real numbers } s_j\} \tag{20a}$$

where the projection operators satisfy

$$\forall i \neq k\colon\; Q_i Q_k = 0 \quad \text{and} \quad Q_1 + \cdots + Q_m = E\,. \tag{20b}$$

This and other structure theorems for full matrix groups shall be used without proof: There is a one-to-one relation between a full group and the matrix algebra generated by this group. The classification of these algebras goes back to WEDDERBURN. After these side remarks, let us go back to (20a) and (20b). A direct and simple calculation yields

$$A^G = Q_1 A Q_1 + \cdots + Q_m A Q_m\,. \tag{20c}$$

At this place we ought to mention an important physical interpretation, due to VON NEUMANN, of the operation (20c): In case A is a density matrix describing a physical state of a quantum statistical system, the operation $A \to A^G$ tells us what happened with the state in the course of a certain physical measurement. The measurement, in our case, is asking m compatible "yes-no-questions", $Q_1, \ldots, Q_m$. The new state the system attains after the measurement is A^G. Since the operations (15) are doubly stochastic, the measurements drive the system to more and more chaotic states. ▮

For arbitrary full G we first look at G_Z, the centre of G. G_Z is a full group, and let (20) now describe the action of G_Z. We define the subgroups G_j of G by

$$G_j = \{U \in G\colon\; Q_j U Q_j = U \text{ and } Q_i U Q_i = E \text{ for } i \neq j\}\,. \tag{21a}$$

From the fullness of G one is allowed to conclude the product decomposition with mutually commuting subgroups

$$G = G_1 G_2 \cdots G_m\,. \tag{21b}$$

With the aid of the corollary of Lemma 2-4 we deduce

$$A^G = \Big(\big(\ldots\big((A^{G_1})^{G_2}\big)\ldots\big)^{G_m}\Big) \quad \text{and} \quad A^G = (A^{G_Z})^G\,.$$

Combining both relations we get

$$A^G = \sum Q_j A^{G_j} Q_j = \sum (Q_j A Q_j)^{G_j} . \tag{21 c}$$

Restricted to the Hilbert subspace which is the range of Q_j, the centre of the group G_j acts as a multiplication by complex numbers. Furthermore, G_j is a full group if considered as a matrix group acting on the range of Q_j. Hence (21) reduces the general case to that one treated below. ▮

The last step in the classification of the maps (15) deals with a full matrix group, G, the centre of which consists of all matrices λE, $|\lambda| = 1$. Let the matrices act on the Hilbert space $\boldsymbol{l}_n^2$. There are naturals, n_1 and n_2, with $n = n_1 n_2$, and the elements of G are given by direct (Kronecker) products

$$G = \{U: E_{n_1} \otimes V, V \in \mathcal{U}(n_2)\} \tag{22 a}$$

where $\mathcal{U}(n_2)$ is the group of all unitary n_2-dimensional matrices. Every matrix $A \in \boldsymbol{B}_n^1$ may be written in various ways in the form

$$A = \sum B_i \otimes C_i \quad \text{with} \quad B_j \in \boldsymbol{B}_{n_j}^1 , \qquad j = 1, 2 . \tag{22 b}$$

and our problem is reduced to a calculation of (15) for the group $\mathcal{U}$ of all unitary matrices of a given dimension. Condition (ii) of Lemma 2-3 shows that C, integrated according to (15) over $\mathcal{U}$, has to be a multiple of E. The factor in front of E has to be the trace of C, for (15) is trace preserving and E goes into E under (15). These reasonings imply with (22 b)

$$A^G = \sum (B_i \otimes E_{n_2}) \operatorname{Tr.} C_i . \tag{22 c}$$

(Remind: Every C_j is a n_2-dimensional matrix, the trace of which is uniquely defined.)

There is again an important physical interpretation of (22 c). At first we write instead of (22 c)

$$A^G = (n_2)^{-1} B \otimes E_{n_2} , \qquad B \in \boldsymbol{B}_{n_1}^1 , \tag{22 d}$$

and assume that A denote a density matrix which is characterizing a certain state of a physical system. In the case at hand there is associated with G a direct product decomposition of the underlying Hilbert space, and we refer now to the pointwise invariant under the action of the G factor. It distinguishes a subsystem of the system. All what is possible to know from the state of the total system by looking only at that subsystem, can be read from B as given by (22 d). B is the density matrix of the state of the subsystem in question, whence the state of the total system is given by A. In the physical literature, B is referred to as a *"reduced density matrix"*.

According to the twofold way matrices enter into quantum statistics, there is, however, yet another interpretation of (15) if written in the form (22 d): If we think of A as not representing a state but an observable quantity, then $A \to B$ as given by (22 d) is called a (special case of a) *"conditional expectation"*. ▮

The mappings considered up to now do not exhaust the doubly stochastic ones. As already has been said, no effective procedure is known for writing them down explicitly. Let us, however, describe a further known class of doubly stochastic maps.

Let us choose matrices $B_1, \dots, B_m$ satisfying

$$\forall i: B_i \geqq 0 \quad \text{and} \quad B_1 + B_2 + \dots + B_m = E \tag{23}$$

and matrices $C_1, \dots, C_m$ with

$$\forall j: C_j \geqq 0 \quad \text{and} \quad \text{Tr. } C_j = 1 . \tag{24}$$

Then the map $\boldsymbol{T}$, defined by

$$A \to \boldsymbol{T}A := \sum (\text{Tr. } AB_j)\, C_j \tag{25}$$

is stochastic. In order that such a map is of the form (13) with (14a) one has to require $\forall i$: $B_iC_i = C_iB_i$. Generally the maps (25) are not completely positive ones.

Hence we get a lot of new examples of doubly stochastic maps by requiring in addition to (23) and (24) the condition

$$E = \sum (\text{Tr. } B_j)\, C_j . \tag{26}$$

We remark on the impossibility of representing the identity map in the form (23)—(26). The identity map cannot be approximated by these transformations either. This prevents us from directly extending 1.9. to the non-commutative case. ∎

Finally we mention that the performance of the transposed matrix is a doubly stochastic map in $\boldsymbol{B}^1$. See [68, 112].

2.3. Convexity and the relation $\succ$

We discuss some aspects of 1.7. in the matrix case: With the help of Theorem 2-1 we trace back the non-commutative problem to the commutative one.

A set $\boldsymbol{N}$ of Hermitian matrices is called *unitarily invariant* iff $A \in \boldsymbol{N}$ implies $UAU^{-1} \in \boldsymbol{N}$ for all unitary matrices U.

In a set $\boldsymbol{N}$ of matrices we distinguish its *diagonal part*

$$\text{diag. } \boldsymbol{N} := \{A \in \boldsymbol{N}\colon A \text{ is diagonal}\} .$$

Clearly, diag. $\boldsymbol{N}$ is a subset of diag. $\boldsymbol{B}^1$ and the latter set we think of as being canonically identified with $\boldsymbol{l}^1$ by associating the element $\{a_{11}, a_{22}, \dots, a_{nn}\}$ of $\boldsymbol{l}^1$ to every diagonal matrix $A = (a_{ik})$. Thus we call diag. $\boldsymbol{N}$ convex (resp. symmetric) iff this is true after identifying diag. $\boldsymbol{N}$ as a subset of $\boldsymbol{l}^1$ as indicated above.

If $\boldsymbol{N}$ is unitarily invariant, then diag. $\boldsymbol{N}$ is symmetric. If $\boldsymbol{N}$ is unitarily invariant and if diag. $\boldsymbol{N}$ is convex, then diag. $\boldsymbol{N}$ is a symmetric convex set. If then $B \succ A$ and $A \in \boldsymbol{N}$ we find a diagonal A' with $A' \sim A$. Because diag. $\boldsymbol{N}$ is symmetric and convex it contains with A' every diagonal matrix B' with $B' \succ A'$ (Theorem 1-21) in the sense of $\boldsymbol{l}^1$ and hence in the sense of Definition 2-2 because of Theorem 2-1. Choosing now the diagonal matrix B' to be equivalent with B: $B' \sim B$ we find $B \in \boldsymbol{N}$. Let us state this as a lemma.

Lemma 2-5. *Let* $\boldsymbol{N}$ *be unitarily invariant set of Hermitian matrices and let* diag. $\boldsymbol{N}$ *be convex. Then* $\succ$ *as defined according to Definition* 2-2 *and according to Definition* 1-2 (*with canonic identification as subset of* $\boldsymbol{l}^1$) *coincides on* diag. $\boldsymbol{N}$. *From* $A \in \boldsymbol{N}$ *and* $B \succ A$ *it follows* $B \in \boldsymbol{N}$.

Corollary. *Let* $\boldsymbol{N}$ *be unitarily invariant and* diag. $\boldsymbol{N}$ *convex. Let* $A \to F(A)$ *be a real unitarily invariant function,*

$$F(A) = F(UAU^{-1}) \quad \text{if } U \text{ is unitary},$$

defined on $\boldsymbol{N}$. *If the restriction of* F *onto* $\boldsymbol{N}$ *is convex, then* $F(B) \leqq F(A)$ *whenever* $B \succ A$ *for two elements of* $\boldsymbol{N}$.

Indeed, by unitary invariance the assertion can be reduced to diag. $\boldsymbol{N}$. But then it is a consequence of Theorem 1-21. ▮

It would be of course not more difficult to prove the statements above without referring to Chapter 1. However, in the reasoning above we indicate in what sense many results of Chapter 1 can be read off from the 'diagonal case' of Chapter 2 and vice versa. The two next theorems will make this even more transparent.

Theorem 2-6. *A unitarily invariant set* $\boldsymbol{N}$ *of Hermitian matrices is a convex set iff* diag. $\boldsymbol{N}$ *is convex.*

This theorem reduces the problem of describing all unitarily invariant sets of Hermitian matrices to the one of finding all symmetric convex subsets of $\boldsymbol{l}^1$. It gives a one-to-one correspondence between these two families of sets. (See [96].)

Theorem 2-7. *Let* $A \to F(A)$ *be a unitarily invariant real function defined on a unitarily invariant convex set* $\boldsymbol{N}$ *of Hermitian matrices.* F *is convex iff its restriction onto* diag. $\boldsymbol{N}$ *is convex.*

This theorem reduces the question of whether a given unitarily invariant function is a convex one to the examination of its behaviour on diagonal matrices. Both theorems gives rise to the following construction. (See [96].)

Construction. Let $\mathcal{N} \subseteqq \boldsymbol{l}^1$ be a convex symmetric set and $\boldsymbol{a} \to f(\boldsymbol{a})$, $\boldsymbol{a} \in \mathcal{N}$, a symmetric convex function. Define

$$\boldsymbol{N} := \{A \text{ Hermitian: Spec } A \subseteqq \mathcal{N}\}.$$

Then $\boldsymbol{N}$ is a convex and unitarily invariant set. (The convexity follows from Theorem 2-6.) Define for all $A \in \boldsymbol{N}$

$$F(A) := f(\text{Spec } A).$$

Then $A \to F(A)$ is a unitarily invariant and convex function on $\boldsymbol{N}$. (This follows from Theorem 2-7.) ▮

As an exercise the reader may analyse the functions $e_s(A)$ and $S_g(A)$ with convex g introduced by the equations (1.27) and (1.37).

Proof of the theorems. Assume $A, B \in \boldsymbol{N}$ and $0 < t < 1$ and consider $C = tA + (1 - t)B$. Eventually after the application of a unitary transformation we are allowed to restrict ourselves to the case where C is diagonal. We need one-dimensional projection operators $P_1, \ldots, P_n$ with $P_iP_j = 0$ for $i \neq j$ and $P_1 + \cdots + P_n = E$ such that $C = \sum c_iP_i$ is a spectral resolution of C. By our argument above we may choose these projection operators diagonal. Now

$$X \to \boldsymbol{T}X := \sum P_iXP_i$$

is of the form (13), (14) and hence chaos-enhancing, i.e. $\boldsymbol{T}X \succ X$ for all Hermitian X. Especially, $\boldsymbol{T}C = C$ by construction. Hence

$$C = \boldsymbol{T}C = t\boldsymbol{T}A + (1 - t)\,\boldsymbol{T}B$$

and $\boldsymbol{T}A$ and $\boldsymbol{T}B$ are both diagonal. Since $A, B \in \boldsymbol{N}$, Lemma 2-5 gives $\boldsymbol{T}A \in \boldsymbol{N}$ and $\boldsymbol{T}B \in \boldsymbol{N}$. Now diag. $\boldsymbol{N}$ is convex and thus $C \in$ diag. $\boldsymbol{N} \subset \boldsymbol{N}$. This proves Theorem 2-6. Next within diag. $\boldsymbol{N}$ we have

$$F(C) \leqq tF(\boldsymbol{T}A) + (1 - t)\,F(\boldsymbol{T}B)$$

according to the assumptions on F. The corollary to Lemma 2-5 ensures us with $F(\boldsymbol{T}A) \leqq F(A)$ for $\boldsymbol{T}A \succ A$. The same is with B. So, finally, we get

$$F(C) \leqq tF(A) + (1 - t)\,F(B)\,,$$

i.e. F is convex. (See [96].) ∎

2.4. Another partial order

We now overcome the restriction of hermiticity for the matrices involved by slightly weakening $\succ$ outside Ω^q.

Definition 2-3. Let $A, B \in \boldsymbol{B}^1$. We write

$$B \rhd A$$

iff B is in the convex hull of the set

$$\{UAV\colon U, V \text{ unitary}\}\,. \tag{27}$$

Being in a finite-dimensional space, the convex hull of a compact set is compact (Bourbaki). The set (27) is compact, and we therefore may already state $B \rhd A$ if B is a limit of finite convex linear combinations out of UAV with unitary U and V. In other words, the set $\{B\colon B \rhd A\}$ is a compact one. The following assertions are immediately clear:

$$C \rhd B\,, \quad B \rhd A\,, \quad \text{gives} \quad C \rhd A\,. \tag{28a}$$

$$B \rhd A \quad \text{implies} \quad B^* \rhd A^*\,. \tag{28b}$$

$$B \rhd A \quad \text{implies} \quad \lambda B \rhd \lambda A \text{ for complex numbers } \lambda. \tag{28c}$$

$$\text{If} \quad B \succ A \quad \text{then} \quad B \rhd A \quad \text{for Hermitian } A, B. \tag{28d}$$

We shall use the notation

$$|A| := (A^*A)^{1/2}\,.$$

With the aid of the polar decomposition we see at once the validity of $|A| \rhd A$ and of $A \rhd |A|$. For convenience we also introduce an equivalence sign for $\rhd$.

Definition 2-3a. We write $A \sim B$ if and only if we have $A \rhd B$ together with $B \rhd A$.

We note

$$|A| \sim A\,. \tag{28e}$$

We need to define the following Ky Fan type functionals

$$\tilde{k}(X, A) = \sup_{U,V} \text{Real Tr. } (UXVA) \tag{29}$$

where the supremum is to run through all unitary matrices U, V. Because of its representation in (29) as a supremum

$$\forall X\colon A \to \tilde{k}(X, A) \quad \text{is convex.} \tag{30}$$

Using the definition, the group properties of the unitary matrices, and the polar decompositon we arrive at

$$\forall U, V \in \mathcal{U}\colon \tilde{k}(X, A) = \tilde{k}(X, UAV)\,, \tag{31 a}$$

$$\tilde{k}(X, A) = \tilde{k}(X, |A|) = \tilde{k}(|X|, A) = \tilde{k}(|X|, |A|)\,. \tag{31 b}$$

Theorem 2-8. *The following statements are mutually equivalent:*

(i) $B \rhd A$.

(ii) $\forall X\colon \tilde{k}(X, B) \leqq \tilde{k}(X, A)$.

(iii) *B is a convex linear combination of elements of the form YAZ where $Y^*Y \leqq E$ and $Z^*Z \leqq E$.*

(iv) $|B| \rhd |A|$.

(v) $\forall j\colon e_j(|B|) \leqq e_j(|A|)$.

Corollary. *Let $A, B \in \Omega^q$, i.e. positive semidefinite matrices with trace equal to one, then $B \rhd A$ if and only if $B \succ A$.*

Proof of the corollary. One direction is clear from (28d). The other one follows from Theorem 2-8, assertion (v): We get $e_j(B) \leqq e_j(A)$ and Tr. B = Tr. A was supposed. Hence, Spec B $\succ$ Spec A which is $B \succ A$, according to Theorem 2-2. ∎

Before we show the validity of Theorem 2-8 we do something to prepare this which is of independent use.

Let us consider the set of diagonal matrices having non-increasingly ordered and non-negative diagonal elements:

$$\boldsymbol{mD}_+ := \{A = (a_{ik})\colon a_{ik} = 0 \text{ for } i \neq k;\ a_{11} \geqq \cdots \geqq a_{nn} \geqq 0\}\,. \tag{32}$$

This set is a representation set for $\rhd$. Indeed, for given matrix C there is just one $C' \in \boldsymbol{mD}_+$ with $C \sim C'$. The diagonal entries of C' are the eigenvalues of $|C|$ decreasingly ordered. A further important property of $\boldsymbol{mD}_+$ is in the fact that it is a multiplicatively closed convex cone, i.e.

$$\begin{gathered} A, B \in \boldsymbol{mD}_+ \quad \text{imply} \quad A + B \in \boldsymbol{mD}_+ \quad \text{and} \quad AB \in \boldsymbol{mD}_+ \\ \text{and} \quad tA \in \boldsymbol{mD}_+ \quad \text{for } t \geqq 0\,. \end{gathered} \tag{33}$$

Finally, the notation of $\boldsymbol{mD}_+$ is good in calculating $\tilde{k}$:

Lemma 2-9. *Let A' and B' be any two matrices. With their representatives $A, B \in \boldsymbol{mD}_+$, $A' \sim A$, $B' \sim B$, we have*

$$\tilde{k}(A', B') = \sup_{U,V} |\text{Tr. } (A'UB'V)| = \text{Tr. } (AB)\,. \tag{34}$$

Proof. Evidently $\tilde{k}(A', B') \geqq \text{Tr. } AB$ and $\tilde{k}(A', B')$ is smaller than the supremum in (34), which runs through all unitaries U, V. Thus we need $|\text{Tr. } (A'UB'V)| \leqq \text{Tr. } AB$. Because U, V vary freely within the unitaries we can choose any pair of matrices equivalent to A', B', i.e. we can take A, B, and prove $|\text{Tr. } (AUBV)| \leqq \text{Tr. } AB$. Now $\beta(X, Y) := \text{Tr. } (AXBY^*)$ is a positive semidefinite scalar product, for A, B are positive semidefinite. Hence $|\beta(X, Y)|$ is dominated by $\{\beta(X, X) \cdot \beta(Y, Y)\}^{1/2}$. Therefore we need only to prove $\text{Tr. } AUBU^{-1} \leqq \text{Tr. } AB$. Let $a_{11} \geqq a_{22} \geqq \cdots$ and $b_{11} \geqq b_{22} \geqq \cdots$ denote the diagonal elements of A and B respectively while the entries of U are denoted by u_{ik}. Then our inequality reads

$$\sum a_{ii} b_{kk} |u_{ik}|^2 \leqq \sum a_{ii} b_{ii} .$$

For U is unitary, the sum of every column and of every row of the matrix $i, k \to |u_{ik}|^2$ equals one. This matrix is therefore a doubly stochastic one by Theorem 1-8. Hence $\sum a_{ii} b_{kk} |u_{ik}|^2$ can be written as a convex linear combination of numbers $\sum a_{ii} b_{k(i),k(i)}$ where $i \to k(i)$ is a permutation. Because of $a_{11} \geqq a_{22} \geqq \cdots$ and $b_{11} \geqq b_{22} \geqq \cdots$ every one of these numbers is smaller than $a_{11}b_{11} + a_{22}b_{22} + \cdots + a_{nn}b_{nn}$. ∎

Proof of Theorem 2-8. The equivalence (i) ⇔ (ii) ⇔ (iii) is a special case of the general Theorem 3-3 which we shall prove independently, and of the compactness of the set $\{B: B \rhd A\}$ stated after Definition 2-3. (i) ⇔ (iv) comes at once from (29e). It remains (ii) ⇔ (v). If Q is a j-dimensional orthoprojection, (34) tells us

$$\forall j\text{-dim. projections } Q\colon\ \tilde{k}(Q, A) = e_j(|A|) . \tag{35}$$

Thus (ii) ⇒ (v). On the other hand, with $X, A \in \boldsymbol{mD}_+$, which suffices to be considered, Lemma 2-9 gives

$$\tilde{k}(X, A) = \text{Tr. } XA = \sum x_{ii} a_{ii} .$$

The last sum can be rewritten, setting $x_{n+1,n+1} = 0$, as

$$\begin{aligned} \sum x_{ii} a_{ii} &= a_{11}(x_{11} - x_{22}) + (a_{11} + a_{22})(x_{22} - x_{33}) + \cdots \\ &= \sum e_j(A)(x_{jj} - x_{j+1,j+1}) . \end{aligned} \tag{36}$$

Hence, generally, we can write

$$\begin{aligned} &\tilde{k}(X, A) = \sum \lambda_j e_j(|A|) , \quad \text{with} \quad \lambda_j := x_{jj} - x_{j+1,j+1} , \\ &\text{and} \quad x_{ii} = 0 \quad \text{for} \quad i > n , \quad \text{and} \quad x_{11} \geqq x_{22} \geqq \cdots \end{aligned} \tag{37}$$

denoting the eigenvalues of $|X|$, decreasingly ordered.

This clearly gives (v) → (ii), and we are done. ∎

Remark. Let A be a matrix. The eigenvalues of $|A|$, ordered decreasingly, are often called the *singular numbers* of A.

There are many inequalities with singular numbers (see [18, 21, 38, 48, 65]). We shall address ourselves only to some typical examples connected with the partial order introduced.

As a first example remember the definition of the functionals $\hat{e}_t(A)$ for Hermitian A, a special case of a functional S_f with concave f, equation (1.43). Further, (1.17) and (1.28) show $\forall s \geqq 0$: $e_s(|B|) \leqq e_s(|A|)$ if statement (v) of Theorem 2-8 is valid. Hence (1.44) and (1.45) give:

Lemma 2-10. *$B \rhd A$ if and only if for all $t \geqq 0$ we have*

$$\hat{e}_t(|B|) \geqq \hat{e}_t(|A|) .$$

We need not consider $t < 0$ for in this domain the inequalities above reduce to $\mathrm{Tr.}\,|B| \leqq \mathrm{Tr.}\,|A|$.

Up to a linear function every concave function $s \to f(s)$ can be approximated by positive linear combinations of the functions $s \to g_t(s)$ defined by equations (1.43a). Using this one can derive the statement of the next but one lemma with the aid of Lemma 2-10. We shall, however, prefer another way which is in rewriting Lemma 1-15 within $\boldsymbol{mD}_+$.

Lemma 2-11 (rewritten Lemma 1-15). *Assume $B \rhd A$ with $A, B \in \boldsymbol{mD}_+$. There is $C \in \boldsymbol{mD}_+$ satisfying*

$$B \leqq C \quad \text{and} \quad C \succ A .$$

If only $e_j(B) \leqq e_j(A)$ for $1 \leqq j \leqq k$, then there is $C \in \boldsymbol{mD}_+$ with $b_{jj} \leqq c_{jj}$ for $1 \leqq j \leqq k$ and

$$\{c_{11}, \ldots, c_{kk}\} \succ \{a_{11}, \ldots, a_{kk}\}$$

in the sense of $\boldsymbol{l}^1_k$.

The following statement, the $\boldsymbol{l}^1$ variant of which we did not mention in Chapter 1 is a result of Weyl, Hardy, Littlewood, and Pólya (combined with Theorem 2-8). (See [38, 66, 111]).

Lemma 2-12. *Denote by $a_1, \ldots, a_k$ and by $b_1, \ldots, b_k$ respectively the first k singular numbers of the arbitrary matrices A and B. Let $s \to f(s)$ be a function defined on $0 \leqq s$ which is convex and monotonously increasing. Then (with $1 \leqq k \leqq n$)*

$$A \rhd B \text{ implies } \sum_{i=1}^{i=k} f(b_i) \leqq \sum_{i=1}^{i=k} f(a_i) .$$

Proof. We have $e_j(|B|) \leqq e_j(|A|)$ for $1 \leqq j \leqq k$ and Lemma 2-11 applies with $a_i = a_{ii}$, $b_i = b_{ii}$. Then $f(c_{11}) + \cdots + f(c_{kk})$ is smaller than $f(a_{11}) + \cdots + f(a_{kk})$ by $\{c_{11}, \ldots\} \succ \{a_1, \ldots\}$ and Theorem 1-21. But monotonicity of f implies $f(b_j) \leqq f(c_{jj})$ for $1 \leqq j \leqq k$. ∎

A slight extension of the statement above is possible with the help of the remark after Theorem 1-22.

A direct consequence of Definition 2-3 reads as follows.

Theorem 2-14. *Let $A \to F(A)$ be a real function defined on $\boldsymbol{B}^1$. If F is convex and $F(A) = F(UAV)$ for all A and all unitaries U, V, then*

$$B \rhd A \quad \text{implies} \quad F(B) \leqq F(A) .$$

The proof is evident.

Let us consider F as supposed by the theorem. If $X, Y \in \boldsymbol{mD}_+$ we have with $X \leqq Y$ also $X \rhd Y$ (Theorem 2-8, v). Hence the restriction onto $\boldsymbol{mD}_+$ of F has to be not only convex, but in addition monotonously increasing. The following construction therefore gives *all* convex $A \to F(A)$ which depend on the singular numbers only.

Theorem 2-15. *Let $\boldsymbol{a} \to f(\boldsymbol{a})$ be defined on the subset of $\boldsymbol{l}^1$*

$$\{\boldsymbol{a} \in \boldsymbol{l}^1 : a_1 \geqq a_2 \geqq \cdots \geqq a_n \geqq 0\} . \tag{38}$$

Let $\boldsymbol{x} = \{x_1 \geqq x_2 \geqq \cdots \geqq x_n\}$ denote the vector of the singular numbers of a given matrix X and define

$$F(X) := f(\boldsymbol{x}) \tag{39a}$$

so that $F(UAV) = F(A)$ for all A and all unitaries U, V

$$X \to F(X) , \qquad X \in \boldsymbol{B}^1 , \tag{39b}$$

is convex if and only if the following two conditions are true:

(i) $\boldsymbol{a} \to f(\boldsymbol{a})$ *is convex on the set* (38),

(ii) $\boldsymbol{a} \leqq \boldsymbol{b}$ *implies* $f(\boldsymbol{a}) \leqq f(\boldsymbol{b})$ *for* $\boldsymbol{a}, \boldsymbol{b}$, *in* (38).

Proof. We already know the necessity of (i) and (ii), and we now assume (i) and (ii) to be true. We choose $0 \leqq t \leqq 1$ and two matrices X, Y, to form $Z = tX + (1-t)\,Y$. Consider $X', Y', Z' \in \boldsymbol{mD}_+$ which are $\sim$-equivalent to X, Y, Z. We first note

$$Z' \rhd tX' + (1-t)\,Y' . \tag{$*$}$$

Indeed, for all $A \in \boldsymbol{mD}_+$ we have by (30)

$$\tilde{k}(A, Z') = \tilde{k}(A, Z) \leqq t\tilde{k}(A, X) + (1-t)\,\tilde{k}(A, Y) .$$

Now $\tilde{k}(A, X) = \tilde{k}(A, X') = \mathrm{Tr.}\, AX'$ by Lemma 2-9 and the same with Y. Hence $\tilde{k}(A, Z') \leqq \tilde{k}\big(A, tX' + (1-t)\,Y'\big)$. Since $A \in \boldsymbol{mD}_+$ has been chosen freely, we arrive at ($*$).

According to Lemma 2-11 there is $W \in \boldsymbol{mD}_+$ with

$$Z' \leqq W \succ tX' + (1-t)\,Y' .$$

The singular numbers of Z', W, X', Y' are just their eigenvalues and the assumption on f now implies $F(Z') \leqq F(W)$ by (ii),

$$F(W) \leqq F\big(tX' + (1-t)\,Y'\big)$$

by (i) and Theorem 1-21, and

$$F\big(tX' + (1-t)\,Y'\big) \leqq tF(X') + (1-t)\,F(Y')$$

by (i). Finally $F(Z) = F(Z')$, $F(X) = F(X')$, $F(Y) = F(Y')$ gives the inequality $F(Z) \leqq tF(X) + (1-t)\,F(Y)$ stated by the theorem. ∎

Example. "Symmetric" seminorms. Let $f = f(\boldsymbol{a})$ be an isometrically invariant real function on $\boldsymbol{l}^1$ with

(j) $f \geqq 0$,

(jj) $f(t\boldsymbol{a}) = tf(\boldsymbol{a})$ for real $t \geqq 0$,

(jjj) $f(\boldsymbol{a} + \boldsymbol{b}) \leqq f(\boldsymbol{a}) + f(\boldsymbol{b})$.

Then f is clearly convex and by Theorem 1-22 we see its monotonicity: $\boldsymbol{a} \leqq \boldsymbol{b}$ implies $\forall j\colon e_j(\boldsymbol{a}) \leqq e_j(\boldsymbol{b})$ and this gives to us $f(\boldsymbol{a}) \leqq f(\boldsymbol{b})$. Thus the restriction onto the set (38) satisfies (i) and (ii) of Theorem 2-15. The corresponding

$$X \to F(X) := f(\boldsymbol{x})$$

where $\boldsymbol{x}$ denotes the vector of the singular numbers of X, is convex on $\boldsymbol{B}^1$. By (j) and (jj) we can see now that F is a *seminorm* on the linear space of matrices. All seminorms with $F(X) = F(UXV)$ for all X and all unitaries U, V, arise in this way, and there exists exactly one such seminorm for every isometrically invariant function f on $\boldsymbol{l}^1$ satisfying (j) to (jjj). Essentially, this result is due to VON NEUMANN [61]. ∎

2.5. Further examples: $A^k \rhd |A|^k$ and related inequalities

WEYL discovered the inequality ($m = 1, 2, 3, \ldots$)

$$\mathrm{Tr.}\, A^m(A^m)^* \leqq \mathrm{Tr.}\, (AA^*)^m .$$

KY FAN generalized it to

$$\forall j\colon e_j(A^m(A^m)^*) \leqq e_j((AA^*)^m) \tag{40}$$

which, indeed, reads in our notation

$$A^m(A^m)^* \rhd (AA^*)^m . \tag{40a}$$

The relation

$$A^k \rhd |A|^k \tag{41}$$

is a little bit different from (40): Using $A^*A \sim AA^*$ it means

$$(A^m(A^m)^*)^{1/2} \rhd (AA^*)^{m/2} . \tag{41a}$$

Both relations, (40) and (41), are true. We now investigate systematically a class of inequalities to which both belong.

Theorem 2-16. *Let A, B be two matrices and choose $A' \sim A$ and $B' \sim B$ with $A', B' \in \boldsymbol{mD}_+$. Then*

$$A + B \rhd A' + B' , \tag{42}$$

$$AB \rhd A'B' . \tag{43}$$

Before proving the theorem we discuss some of its implications. Having an expression

$$X = A_1A_2 \cdots A_k + B_1B_2 \cdots B_j + \cdots \tag{44a}$$

with certain matrices $A_1, \ldots$ we associate another matrix with this expression as follows. We choose matrices $A_1', \ldots, B_1', \ldots$ out of $\boldsymbol{mD}_+$ which are correspondingly equivalent: $A_1' \sim A_1, \ldots, B_1' \sim B_1, \ldots$ and perform

$$Y = A_1'A_2' \cdots A_k' + B_1'B_2' \cdots B_j' + \cdots . \tag{44b}$$

Evidently, by Theorem 2-16, we may conclude

$$X \rhd Y . \tag{44c}$$

As special case of this procedure, $A_j \sim A$, $j = 1, \ldots, k$, gives

$$A_1A_2 \cdots A_k \rhd |A|^k$$

for $|A|$ is unitarily equivalent to its representative in $\boldsymbol{mD}_+$. Choosing $\forall i$: $A_i = A$ we get (41). Choosing $k = 2m$, $A_i = A$ for $1 \leqq i \leqq m$, $A_i = A^*$ for $m < i \leqq 2m$, we get (40).

Theorem 2-16 and the technique of proving it can be traced back to Horn, Krein, Ky Fan, and Weyl, see [30, 39, 58, 111].

Let us now start with the proof. Choose $D \in \boldsymbol{mD}_+$. Then

$$\tilde{k}(D, A + B) \leqq \tilde{k}(D, A) + \tilde{k}(D, B) = \tilde{k}(D, A') + \tilde{k}(D, B')$$

and the last term equals $\tilde{k}(D, A' + B')$ by Lemma 2-9, equation (34). Hence (42) is established. (43) is connected with interesting aspects which we study in more detail and we stop at this point, postponing the proof of (43).

Let us introduce an auxiliary (for us) partial ordering.

Definition 2-4. Let us denote by $x_1 \geqq x_2 \geqq \cdots$ and $y_1 \geqq y_2 \geqq \cdots$ the singular numbers of X and Y. We write

$$X \vdash Y \tag{45}$$

if and only if

$$\forall\, 1 \leqq j \leqq n: \quad x_1x_2 \cdots x_j \leqq y_1y_2 \cdots y_j$$

and

$$\operatorname{Det} |X| = x_1x_2 \cdots x_n = y_1y_2 \cdots y_n = \operatorname{Det} |Y| .$$

The relevance of this partial order for Golden-Thomson-like inequalities has been stressed by Lenard and Thomson, see [36, 52, 90, 91]. If X^{-1} exists we obviously have

$$X \vdash Y \qquad \text{iff} \quad \ln |X| \succ \ln |Y| . \tag{46}$$

It is now tempting to apply our knowledge concerning $\succ$. One example, to which we restrict ourselves, goes this way.

Let $t \to g(t)$ be convex on the real line. Lemma 1-25 asserts the convexity of $A \to e_j(g(A))$ for Hermitian A. The corollary to Lemma 2-5 (or likewise Theorem 1-21) provides us with

$$X \vdash Y \quad \text{implies} \quad \forall j, \forall \text{ convex } g: \;\; e_j(g(\ln |X|)) \leqq e_j(g(\ln |Y|)) . \tag{47}$$

Choosing $g(t) = \exp t$ we get (see Theorem 2-2, (e))

$$X \vdash Y \quad \text{implies} \quad X \rhd Y . \tag{48}$$

From the definition we have trivially $X \vdash Y$ iff only $|X|^\beta \vdash |Y|^\beta$ for one $\beta > 0$. This sharpens conclusion (48). (We likewise may use the function $g(t) = \exp \beta t$ in (47).) On the other side we get from

$$\sum_{i=1}^{j} (x_i)^\beta \leqq \sum_{i=1}^{j} (y_i)^\beta \quad \text{for} \quad \beta \to 0$$

the validity of $|X|^\beta \rhd |Y|^\beta$ because $\beta^{-1}(x^\beta - 1)$ goes to $\ln x$.

Using continuity arguments to go from the situation $\operatorname{Det} X \neq 0$ to the general one, the reasoning above is summarized in

Lemma 2-17. *$X \vdash Y$ implies $\forall \beta > 0$: $|X|^\beta \rhd |Y|^\beta$. If $|X|^\beta \rhd |Y|^\beta$ is valid for a set of numbers $\beta > 0$, the limit inferior of which is zero, and if* $\operatorname{Det} |X| = \operatorname{Det} |Y|$, *we have $X \vdash Y$.*

The lemma points to a sharpening of (43) to

$$\forall \beta > 0\colon \quad |AB|^\beta \rhd (A'B')^\beta \quad \text{(notation of Lemma 2-4)} \tag{49}$$

which is (43) for $\beta = 1$, and which in turn means with the same notation

$$AB \vdash A'B' \tag{49'}$$

for the determinants of $|AB|$ and $A'B'$ coincide. (49') is (in another notation) due to Horn and Weyl. Let us consider the matrices in question as linear operators acting on l_n^2 with scalar product $(.,.)$. For any set $\xi_1, \dots, \xi_m$, $1 \leqq m \leqq n$, of linearly independent vectors we have the inequality (Weyl)

$$\operatorname{Det} \{(\xi_i, A^*A\xi_k)\,,\ 1 \leqq i, k \leqq m\} \leqq \lambda_m \operatorname{Det} \{(\xi_i, \xi_k)\,,\ 1 \leqq i, k \leqq m\}$$

$$\text{where} \quad \lambda_m = (a_1 a_2 \cdots a_m)^2 \tag{50}$$

with the singular numbers $a_1 \geqq a_2 \geqq \cdots$ of A.

Now we denote by $\xi_1, \dots, \xi_m$ an orthonormal system of eigenvectors for the largest m eigenvalues, $c_1^2, \dots, c_m^2$, of A^*B^*BA. Then applying inequality (50) twice we get

$$(c_1 \cdots c_m)^2 = \operatorname{Det} \{(\xi_i, A^*B^*BA\xi_k)\} \leqq (a_1 \cdots a_m)^2 \operatorname{Det} (\xi_i, B^*B\xi_k)$$
$$\leqq (a_1 \cdots a_m b_1 \cdots b_m)^2 .$$

This is the proof of (49') because the squared singular numbers of $|AB|$ coincide with the eigenvalues of A^*B^*BA.

Finally, we show (50). There is a complete orthonormal system $\eta_1, \dots, \eta_n$ satisfying $\forall j$: $A^*A\eta_j = a_j^2\eta_j$. Introducing the m-by-n-matrices ($1 \leqq i \leqq m$, $1 \leqq j \leqq n$)

$$M = \{(\xi_i, \eta_j)\} \quad \text{and} \quad N = \{(\xi_i, \eta_j)\, a_j\}\,,$$

we rewrite (50) as

$$\operatorname{Det} (NN^*) \leqq (a_1 \cdots a_m)^2 \operatorname{Det} (MM^*)\,.$$

We associate, with every vector of integers $\boldsymbol{r} = \{r_1, \dots, r_m\}$ with $1 \leqq r_1 \leqq r_2 \leqq \cdots \leqq r_m \leqq n$, the m-by-m-matrices $M_{\boldsymbol{r}}$ and $N_{\boldsymbol{r}}$ consisting of the columns of M and of N respectively indexed by $r_1, \dots, r_m$. Then the desired inequality reads

$$\sum \operatorname{Det} (N_{\boldsymbol{r}} N_{\boldsymbol{r}}^*) \leqq (a_1 \cdots a_m)^2 \sum \operatorname{Det} (M_{\boldsymbol{r}} M_{\boldsymbol{r}}^*) \tag{*}$$

where the sum is extended over the vectors $\boldsymbol{r}$ of integers as defined above. But $N_r = M_r A_r$, where A_r is the diagonal matrix consisting of the diagonal elements $a_{r_1}, \dots, a_{r_m}$. Hence

$$\mathrm{Det}\,(N_r N_r^*) = \mathrm{Det}\,(A_r)^2\,\mathrm{Det}\,(M_r M_r^*)$$

and

$$\mathrm{Det}\,(A_r)^2 = (a_{r_1} \cdots a_{r_m})^2 \leqq (a_1 \cdots a_m)^2$$

proved (∗) and, therefore, (50). ▮

2.6. Miscellaneous

We collect some further properties of $\succ$ and problems connected with this partial order which seem worthwhile mentioning.

2.6.1. Chacraterization of equivalence classes

Let A be a Hermitian matrix of dimension n. Up to unitary equivalence A is characterized by the n-tupel of real numbers

$$\{e_1(A), e_2(A), \dots, e_n(A)\}\ .$$

By remembering that $e_j(A)$ stands for the sum of the j largest eigenvalues, one finds without difficulty:

Let $\boldsymbol{a} \in \boldsymbol{l}^1$. If and only if

$$a_1 \leqq a_2 \cdots \leqq a_n \tag{51}$$

$$\forall\, 1 < j < n\colon a_{j-1} + a_{j+1} \leqq 2a_j \tag{52}$$

there is a Hermitian A of dimension n such that

$$\forall j\colon e_j(A) = a_j\ .$$

The inequality (52) expresses the concavity of $s \to e_s(A)$ if restricted to the integers $j = 1, \dots, n$ (see Lemma 1-19).

2.6.2. Lattice properties

Let $\boldsymbol{N}$ be a bounded set of Hermitian matrices (dimension n). There is up to unitary equivalence one and only one matrix B with

(i) $B \succ A$ for all $A \in \boldsymbol{N}$.

(ii) From $C \succ A$ for all $A \in \boldsymbol{N}$ it follows $C \succ B$.

We abbreviate B by $\vee\, \boldsymbol{N}$.

Indeed, $b_s = \inf e_s(A)$, where A is running through $\boldsymbol{N}$ is a concave and monotonous function of s. Hence $j \to b_j$ fulfils (52). The condition (51) is obviously fulfilled. Up to equivalence there is now a matrix B with $\forall j\colon b_j = e_j(A)$.

The dual to $\vee$ lattice operation is characterized by the existence of B' such that

(j) $B' \prec A$ for all $A \in \boldsymbol{N}$.

(jj) From $C' \prec A$ for all $A \in \boldsymbol{N}$ it follows $C' \prec B'$.

Iff B' exists we write $B' = \wedge \boldsymbol{N}$. Its existence is seen for bounded sets $\boldsymbol{N}$ of matrices as following. Define

$$\boldsymbol{M} := \{D : D \prec A \text{ for all } A \in \boldsymbol{N}\} .$$

Then the matrix $\vee \boldsymbol{M}$ satisfies (j) and (jjj). Hence $B' = \vee \boldsymbol{M}$.

2.6.3. Kronecker products

In the following $X_1, \dots, X_m, Y_1, \dots, Y_m$ denote n-dimensional matrices. We perform the n^m-dimensional matrices

$$X = X_1 \otimes X_2 \otimes \cdots \otimes X_m , \qquad Y = Y_1 \otimes Y_2 \otimes \cdots \otimes Y_m$$

where $\otimes$ is the sign for the Kronecker product, i.e. the tensor product. We assume all matrices to be Hermitian.

Statement 1: Assume $\forall j: X_j \succ Y_j$. Then $X \succ Y$.

To see this we expand every X_j according to Theorem 2-2, (b). After performing the tensor product we see X expanded in a convex sum of matrices unitarily equivalent with Y. ∎

Statement 2: Define with $\beta \neq 0$: $s \to f(s) = s^\beta$. Then $S_f(X) = S_f(X_1)\, S_f(X_2) \cdots S_f(X_m)$.

Denoting by $x_{j1}, \dots, x_{jn}$, the eigenvalues of X_j we can identify the eigenvalues of X by the set of all products $x_{1i_1} x_{2i_2} \cdots x_{mi_m}$. The assertion follows from the special structure of the function f.

Let us now assume in (53)

$$\forall i: X_i = X' \quad \text{and} \quad Y_i = Y'$$

so that X and Y denote the mth Kronecker power of X' and Y' respectively. Statement 2 implies for the special functions $f(s) = s^\beta$ the simultaneous validity or non-validity of the inequalities $S_f(X) \geqq S_f(Y)$ and $S_f(X') \geqq S_f(Y')$. On the other hand, $X \succ Y$ does *not*, generally, imply $X' \succ Y'$. (A trivial exception is the situation in which $\dim X' = \dim Y' = 2$.)

On the other hand we learn from the argument giving statement 2 the relevance of Definition 2-4, namely

Statement 3: $X \vdash Y$ iff $X' \vdash Y'$.

Using the behaviour of the numbers $e_k(.)$ one easily get for the direct *sums*

Statement 4: $X' \otimes X' \prec Y' \otimes Y'$ iff $X' \prec Y'$.

2.6.4. The use of coherent vectors

Let (X, dx) be a measure space and $x \to P(x)$ a measurable function of X into the one-dimensional orthoprojections of a finite-dimensional (for simplicity) Hilbert space. Assume

$$\int P(x)\, dx = E \quad \text{(identity matrix)} . \tag{54}$$

Then we get $S_f(A)$ by integrating Tr. $P(x)\, f(A)\, dx$ over X. Being a one-dimensional projection we have

$$\text{Tr. } P(x)\, f(A) \geqq f\big(\text{Tr. } P(x)\, A\big)$$

for convex f (see equation (1.42), where f was assumed to be concave).

It follows for all convex f

$$S_f(A) \geqq \int f\big(\text{Tr. } P(x)\, A\big)\, dx\,. \tag{55}$$

Applied to the exponential function this gives an estimation of the free energy in certain cases. This can be seen in [54, 84]. Further, let Y be a measurable subset of X with a measure less than s. Hence we get an operator with a trace less than s by integrating $P(x)\, dx$ over Y and this operator is in addition smaller than E by (54). Applying the defining equation (1.27) we see:

$$e_s(A) \geqq \int_Y \text{Tr. } P(x)\, A\, dx \quad \text{if} \quad s \geqq \int_Y dx\,. \tag{56}$$

A further investigation is possible if

$$A = \int g(x)\, P(x)\, dx\,. \tag{57}$$

Consider the operator

$$C_n := \left[\int \big(1 - g(x)/n\big)\, P(x)\, dx\right]^n.$$

Let ξ be an eigenvector of C_n which is normed. Then $\langle \xi, P(x)\, \xi\rangle\, dx$ is a probability measure and the convexity of the function $s \to s^n$ gives:

$$\begin{aligned}\langle \xi, C_n \xi\rangle &= \left[\int (1 - g/n)\, \langle \xi, P(x)\, \xi\rangle\, dx\right]^n \\ &\leqq \int (1 - g/n)^n\, \langle \xi, P(x)\, \xi\rangle\, dx\,.\end{aligned}$$

Hence we have

$$\text{Tr. } C_n \leqq \int (1 - g/n)^n\, dx\,.$$

Performing the limit $n \to \infty$ we get the inequality

$$\text{Tr. } \exp(-A) \leqq \int \exp\big(-g(x)\big)\, dx\,. \tag{58}$$

For another method of proving this see [54, 84], see also [102, 109].

2.6.5. The problem to order pairs of matrices (see [14])

Let us consider four density matrices A, B, C, D, i.e. non-negative matrices with trace one. Under which circumstances one can find a stochastic transformation $\boldsymbol{T}$ (see Definition 2-2) fulfilling

$$\boldsymbol{T}A = C \quad \text{and} \quad \boldsymbol{T}B = D\,.$$

A necessary condition for the existence of $\boldsymbol{T}$ reads

$$\forall s \geqq 0: \quad \|A - sB\|_1 \geqq \|C - sD\|_1\,. \tag{59}$$

It turns out, see [14], that (59) is even a sufficient condition *if* the dimension of the matrices is *two*. The condition (59) is insufficient, in general, for dimensions higher than 2. It is unknown and probably a deep problem how to get manageable sufficient conditions for the existence of $\boldsymbol{T}$.

3. The order structure in the state space of C^*- and W^*-algebras

3.1. Preliminaries

We are now going to introduce and to investigate a partial ordering in the state space of C^*- and W^*-algebras which generalize the relation $\succ$ of Chapter 2. We, as an example, may ask ourselves how to formulate a variant of Theorem 2-2 appropriately for such algebras. Especially for W^*-algebras we present an answer to these questions which we believe to be the key to this problem.

Let us introduce some notations and simple statements. We refer to [17, 26, 27, 62, 80] as a reference to standard notations and basic facts of the theory of C^*- and W^*-algebras. Proofs, which the reader can find there, are not usually reproduced here.

Let $\mathcal{A}$ be a C^*-algebra. Its elements are usually denoted by $a, b, c, \ldots$ We always assume the existence of a unity element. It will be called $\mathbf{1}$ (or $\mathbf{1}_{\mathcal{A}}$). We shall use the symbols

$$\mathcal{A}_h, \mathcal{A}_+, \mathcal{U}, \mathcal{A}^*, \mathcal{A}_h^*, \mathcal{A}_+^*, \mathcal{S} ,$$

in a way described below.

The *Hermitian part*, $\mathcal{A}_h$, of $\mathcal{A}$ is the real linear space $\{a \in \mathcal{A} : a = a^*\}$. An element $a \in \mathcal{A}$ is called *positive*, and we write $a \geqq 0$ in that case, iff $a = b^*b$ with suitable $b \in \mathcal{A}$. The *positive cone* of $\mathcal{A}$, denoted by $\mathcal{A}_+$, is the set of all positive elements of $\mathcal{A}$. By $\mathcal{U}$ (or by $\mathcal{U}_{\mathcal{A}}$) we denote the *group of unitary elements*, u, of $\mathcal{A}$, i.e. the group of those elements, u, satisfying $uu^* = u^*u = \mathbf{1}$.

For the C^*-norm of an element $a \in \mathcal{A}$ we write $||a||$. The *dual*, $\mathcal{A}^*$, of $\mathcal{A}$ consists of all linear functionals

$$\omega : a \to \omega(a) , \qquad a \in \mathcal{A} ,$$

bounded by this norm. The norm $||\omega||$ of ω is given by

$$||\omega|| = \sup_{a \neq o} |\omega(a)| \cdot (||a||)^{-1} . \tag{1}$$

(This norm corresponds to the $||.||_1$-norm of $\boldsymbol{B}^1$ in Chapter 2.) $\omega \in \mathcal{A}^*$ is *Hermitian* iff $\omega(a^*) = \overline{\omega(a)}$. Here the "bar" denotes the complex conjugate of a complex number.

Next, the *cone of positive linear functionals* is referred to by the letter $\mathcal{A}_+^*$. It consists of those linear functionals, ω, for which $\omega(a) \geqq 0$ for all $a \in \mathcal{A}_+$. The existence of the unity element guarantees $\mathcal{A}_+^* \subseteqq \mathcal{A}_h^*$. Remember $||\omega|| = \omega(\mathbf{1})$ for positive linear functionals.

By $\mathcal{S}$ (or by $\mathcal{S}_{\mathcal{A}}$) we denote the *state space* of $\mathcal{A}$. It consists of all states of the algebra, and a *state* of $\mathcal{A}$ is a positive and normed linear functional, i.e.

$$\mathcal{S} := \{\omega \in \mathcal{A}_+^* : \omega(1) = 1\} . \tag{2}$$

Or, equivalently, a state is a positive linear functional of norm one.

As a matter of fact, the last notation reflects a physical interpretation: It is instructive to consider the Hermitian elements of $\mathcal{A}$ (or a suitable subset of them) as representing "observables" of a physical system, and to identify the states of $\mathcal{A}$ (or a suitable subset of them) as the possible "states" the physical system admits. Then the number $\omega(a)$ is said to be the "expectation value" of the "observable a" in case the physical system is in the "state ω". This interpretation is, in our opinion, of exceptional heuristic value (see [29, 37, 75, 92, 95]).

Let us in this connection shortly consider the meaning of the finite-dimensional normed spaces $\boldsymbol{l}^1$ and $\boldsymbol{B}^1$ of Chapters 1 and 2. Let $\mathcal{A} = C(X)$ and X the set $X = \{1, 2, \ldots, n\}$. Then every $a \in \mathcal{A}$ is a complex valued function $a: \boldsymbol{i} \to a(\boldsymbol{i})$ defined on X. To every element $\boldsymbol{b} \in \boldsymbol{l}^1$ we associate a linear functional by $a \to \omega(a) = b_1 a(1) + \cdots + b_n a(n)$ if $\boldsymbol{b} = \{b_1, \ldots, b_n\}$. This defines an isomorphism of the normed linear space $\boldsymbol{l}^1$ onto the normed linear space $\mathcal{A}_h$. Under this isomorphism a positive element of $\boldsymbol{l}^1$ is mapped on a positive linear functional, and the states of $\mathcal{A}$ are the images of the probability vectors of $\boldsymbol{l}^1$.

A similar construction is with $\boldsymbol{B}^1$: Let $\mathcal{A} = \mathcal{M}_n$ denote the algebra of n-by-n-matrices. If $a = (a_{ik})$ is such a matrix, $||a||^2$ is the largest eigenvalue of a^*a. If now $A \in \boldsymbol{B}^1$ (we denote matrices by different letter according to their different roles) we associate to A a linear functional ω of $\mathcal{M}_n$ defined by $a \to \omega(a) = \text{Tr.}\,(aA)$.

Again, this defines an isometry of the linear normed space $\boldsymbol{B}^1$ onto the dual $\mathcal{M}_n^*$ of $\mathcal{M}_n$. Under this isomorphism ω is positive iff $A \geqq 0$, and ω is a state iff A is a density matrix (see (2.11)). Thus the isomorphism induces an affine isomorphism of Ω^q onto $\mathcal{S}$. (In the case of $\boldsymbol{l}^1$ there is an affine isomorphism of Ω as defined in 1.8. onto the state space of $C(X)$, $X = \{1, 2, \ldots, n\}$.) ▮

In $\mathcal{A}^*$ we shall use not only the norm topology, but the weak topology too. The weak or *w^*-topology* is given by the family of seminorms

$$\omega \to \sum |\omega(a_j)| , \quad \text{finite sum} , \tag{3}$$

which are built with the help of arbitrary finite subsets $a_1, a_2, \ldots$ of elements from $\mathcal{A}$. Every w^*-closed and bounded by the norm (1) subset of $\mathcal{A}$ is w^*-compact. For example the state space, $\mathcal{S}$, is weakly compact.

We need an elementary lemma.

Lemma 3-1. *Let $A: \omega \to A(\omega)$ be a w^*-continuous linear functional defined on A^*. Then there is an element $a \in \mathcal{A}$ with $\forall \omega: A(\omega) = \omega(a)$.*

Proof. Let A be bounded by the seminorm (3). If $\mathcal{N}$ denotes the linear subspace generated by the finitely many $a_j \in \mathcal{A}$ occurring in (3), then $A(\omega) = 0$ if ω vanishes on $\mathcal{N}$. If $b_1, \ldots, b_m$ is a base of $\mathcal{N}$, and if the linear functionals $\omega_1, \ldots, \omega_m$ are such that $\omega_j(b_k) = \delta_{jk}$, we consider the element $a = \sum A(\omega_j)\, b_j$. For every linear functional ω the difference $\omega - \sum \omega(b_k)\, \omega_k$ vanishes on $\mathcal{N}$. This gives $A(\omega) = \sum \omega(b_k)\, A(\omega_k) = \omega(a)$. ▮

3.2. The relation $\succ$; definitions and elementary results

We are now going to introduce the (pre-) partial ordering $\succ$ in the state space $\mathcal{S}$ of a C^*-algebra $\mathcal{A}$. (See [2, 19, 97, 100, 107].) As a matter of fact, there are several possibilities to extend this notation to larger portions of $\mathcal{A}^*$, and we intend to mention two of them below. In the following chapters, however, we are concerned mainly with the properties shown by the relation $\succ$ on the state space $\mathcal{S}$. (Faced with the problem of making a choice, we follow our belief of a possible physical relevance.) Let us start with the following notations.

Notation. With $\omega \in \mathcal{A}^*$ and $a, b \in \mathcal{A}$ we denote by

$$\omega^{a,b} \tag{4a}$$

the linear form

$$\forall d \in \mathcal{A}: \quad d \to \omega(a^*db) \tag{4b}$$

and we abbreviate

$$\omega^a := \omega^{a,a} . \tag{4c}$$

Definition 3-1. For $\varrho, \omega \in \mathcal{A}^*$ we write

$$\varrho \succ \omega$$

iff ϱ is in the w^*-closed convex hull generated by all ω^u with $u \in \mathcal{U}$:

$$\varrho \in \omega^*\text{-closure convex hull } \{\omega^u: u \in \mathcal{U}\} . \tag{5}$$

Every complex-linear space can be canonically considered as a real-linear space. Here and in the following the property of being *convex* is always meant as relative to this *real* linear structure of $\mathcal{A}^*$.

We now use well-known separation theorems for compact convex sets in order to express $\succ$ by means of a set of inequalities. We say for short $\omega \to F(\omega)$ is a *l.s.c. function* iff it is (in its w^*-closed domain of definition) w^*-lower semicontinuous. The supremum of any family of l.s.c. functions is l.s.c. again. A subset of $\mathcal{A}^*$ is called *unitarily invariant* if with ω it contains ω^u for all $u \in \mathcal{U}$. A function which is constant along every orbit $\{\omega^u, u \in \mathcal{U}\}$ intersecting its domain of definition is accordingly called *unitarily invariant.*

A notation very important for us is the following one. For every $\omega \in \mathcal{A}^*$ and every $a \in \mathcal{A}$ we define

$$K(\omega, a) := \{\sup \text{Real } \omega^u(a), \ u \in \mathcal{U}\} . \tag{6}$$

These suprema are called *Ky Fan functionals.* For fixed $a \in \mathcal{A}$ they are, obviously, l.s.c. functions, for every ω^u is l.c.s. The suprema of linear functionals are not only convex but even subadditive:

$$K(\omega + \nu, a) \leqq K(\omega, a) + K(\nu, a) , \tag{7a}$$

$$K(\omega, a + b) \leqq K(\omega, a) + K(\omega, b) . \tag{7b}$$

Further

$$\forall s \geqq 0: \quad K(s\omega, a) = K(\omega, sa) = sK(\omega, a) , \tag{8}$$

$$\forall u \in \mathcal{U}: \quad K(\omega, a) = K(\omega^u, a) = K(\omega, u^{-1}au) . \tag{9}$$

Theorem 3-2. *The following statements are mutually equivalent*:

(i) $\varrho \succ \omega$.

(ii) $\forall a \in \mathcal{A}: \quad K(\varrho, a) \leqq K(\omega, a)$.

(iii) *Let $\mathcal{K}$ be a w^*-closed, convex, and unitarily invariant subset of $\mathcal{A}$ which contains ω. Then $\varrho \in \mathcal{K}$. For every unitarily invariant, convex, l.s.c. function F which is defined on $\mathcal{K}$, it is*

$$F(\varrho) \leqq F(\omega) .$$

Proof. Assume (i). The set (5) is then the intersection of all sets $\mathcal{K}$ mentioned in (iii). Hence $\varrho \in \mathcal{K}$. The set $\{\nu \in \mathcal{K}: F(\nu) \leqq F(\omega)\}$ is convex, unitarily invariant and weakly closed. Hence it contains with ω also ϱ. The step (iii) $\to$ (ii) is trivial. We prove (ii) $\to$ (i) by showing the existence of a Ky Fan functional not obeying (ii) if ϱ is not in the set (5).

The set (5) is weakly compact because it is weakly closed and bounded in norm by $\|\omega\|$. If ϱ is not in this set there exists a w^*-continuous real linear functional, L, such that $\forall \nu \succ \omega: L(\nu) \leqq L(\varrho) + 1$. (Because (5) is convex and w^*-compact.) Define $F(\sigma) = \{\sup L(\sigma^u),\ u \in \mathcal{U}\}$. Then $F(\nu) \leqq L(\varrho) + 1 \leqq F(\varrho) + 1$ for all $\nu \succ \omega$. Hence $F(\omega) < F(\varrho)$. But $\sigma \to L(\sigma) - iL(i\sigma)$ is complex linear and w^*-continuous. By Lemma 3-1 this implies the existence of $a \in \mathcal{A}$ with $L(\sigma) - iL(i\sigma) = \sigma(a)$. This gives $F(\sigma) = K(\sigma, a)$ which is the desired contradiction, i.e. (i) follows from (ii). (See [97].) ▮

Corollary. *Assume $\varrho, \omega \in \mathcal{A}_h^*$, $\varrho(\mathbf{1}) = \omega(\mathbf{1})$, and*

$$\forall a \in \mathcal{A}_+: K(\varrho, a) \leqq K(\omega, a) . \tag{10}$$

Then we have $\varrho \succ \omega$.

Proof. In proving (ii) $\to$ (i) we use only the properties of the restriction on $\mathcal{A}_h^*$ of L in that case. Hence the proof goes through after replacing L by

$$L' : \sigma \to (1/2)\, L(\sigma + \sigma^*) + (1/2)\, L(i\sigma - i\sigma^*) .$$

But then $\sigma \to L'(\sigma) - iL'(i\sigma)$ is real for Hermitian σ and our $a \in \mathcal{A}_h$. This means that we can restrict ourselves to Hermitian elements in (ii) in the case at hand. Now $a + s\mathbf{1} \in \mathcal{A}_+$ for large enough s and

$$K(\sigma, a + s\mathbf{1}) = K(\sigma, a) + s\sigma(\mathbf{1}) \quad \text{if} \quad \sigma \in \mathcal{A}_h . \tag{11}$$

The validity of (ii) for Hermitian $a \in \mathcal{A}$ is now seen to be equivalent with (10) and $\varrho(\mathbf{1}) = \omega(\mathbf{1})$. ▮

Let us supplement Definition 3-1 with the following remarks. We express $\varrho \succ \omega$ verbally by saying "*ϱ is more chaotic than ω*". (Remember the remarks after Definition 2-1!) Immediately from the definition one gets the statements

$$\omega'' \succ \omega' \quad \text{and} \quad \omega' \succ \omega \quad \text{imply} \quad \omega'' \succ \omega , \tag{12}$$

$$\varrho \succ \omega \quad \text{implies} \quad \varrho^* \succ \omega^* , \tag{13}$$

$$\varrho \succ \omega \quad \text{implies for all complex numbers } \lambda \text{ the relation} \tag{14}$$

$$\lambda \cdot \varrho \succ \lambda \cdot \omega .$$

If we denote by $\mathcal{Z}$ the *centre* of $\mathcal{A}$, we have

$$\varrho \succ \omega \quad \text{implies} \quad \varrho(c) = \omega(c) \quad \text{for all} \quad c \in \mathcal{Z} . \tag{15}$$

Our next task is to compare $\succ$ with another partial order in $\mathcal{A}^*$.

Definition 3-2. For every pair $\varrho, \omega \in \mathcal{A}^*$ we write

$$\varrho \rhd \omega$$

iff

$$\varrho \in w^*\text{-closure convex hull} \quad \{\omega^{u,v}: u, v \in \mathcal{U}\} . \tag{16}$$

We see at once that $\varrho \succ \omega$ implies $\varrho \rhd \omega$. Further, the statements (12), (13), and (14) remain true after replacing $\succ$ by $\rhd$. In complete analogy we define

$$\tilde{K}(\omega, a) := \{\sup \operatorname{Real} \omega^{u,v}(a), u \in \mathcal{U}, v \in \mathcal{U}\} \tag{17}$$

such that

$$\tilde{K}(\omega, a) = \sup_{u \in \mathcal{U}} K(\omega, ua) = \sup_{u \in \mathcal{U}} K(\omega, au) . \tag{18}$$

While the Ky Fan functionals remain constant on every orbit $\{\omega^u, u \in \mathcal{U}\}$, the functional $\tilde{K}(\omega, a)$ is constant along every "double orbit" $\{\omega^{u,v}$ with $u, v \in \mathcal{U}\}$.

In the next theorem $\mathcal{E}$ stands for the *unit sphere* of $\mathcal{A}$.

Theorem 3-3. *The following statements are mutually equivalent:*

(i) $\varrho \rhd \omega$.

(ii) $\forall a \in \mathcal{A}: \tilde{K}(\varrho, a) \leqq \tilde{K}(\omega, a)$.

(iii) *For every convex l.s.c. function F, defined on $\mathcal{A}^*$, and satisfying for all σ and*

$$\forall u, v \in \mathcal{U}: F(\sigma) = F(\sigma^{u,v}) \tag{19}$$

it is

$$F(\varrho) \leqq F(\omega) .$$

(iv) $\varrho \in w^*$*-closure convex hull* $\{\omega^{a,b}$ *with* $a, b \in \mathcal{E}\}$.

Proof. The equivalence of (i), (ii), and (iii) is obtained straightforwardly through the same arguments as in proving that Theorem 3-2. (i) → (iv) is trivial. (iv) → (i) is equivalent with $\omega^{a,b} \rhd \omega$ for all $a, b \in \mathcal{E}$. The latter assertion follows immediately from a theorem due to Russo and Dye [28, 40, 76] which reads

$$\mathcal{E} = \text{norm-closure convex hull } \mathcal{U} . \tag{20}$$

As a consequence of (20), $\omega^{a,b}$ is even in the norm-closure of the convex hull generated by all $\omega^{u,v}$ with $u, v \in \mathcal{U}$. ∎

We now aim to relate Theorems 3-2 and 3-3.

For $a \in \mathcal{A}_+$ and $\omega \in \mathcal{A}_+^*$ the Hermitian form

$$c, d \to \omega(c^*ad)$$

is positive semidefinite. Hence

$$|\omega(c^*ad)|^2 \leqq \omega(c^*ac) \cdot \omega(d^*ad) .$$

Assuming $c, d \in$ convex hull $\mathcal{U}$ we find convex linear combinations $c = \sum s_i u_i$, $d = \sum t_i v_i$ with unitaries u_i, v_j. Then

$$|\omega(c^*ad)| \leqq \sum_{i,j} s_i t_i\, \omega(u_i a u_i)^{1/2} \cdot \omega(v_j a v_j)^{1/2} .$$

By (6) we have $\omega(u^*au)$ bounded by $K(\omega, a)$ for all $u \in \mathcal{U}$. Therefore, $|\omega(c^*ad)| \leqq K(\omega, a)$. The theorem of Russo and Dye, (20), ensures us of the validity of this inequality for all $c, d \in \mathcal{E}$. This gives

$$\tilde{K}(\omega, a) \leqq \sup |\omega(c^*ad)| \leqq K(\omega, a)$$

where the supremum runs through all $c, d \in \mathcal{E}$. The opposite inequality is obvious. The result reads

Lemma 3-4. *For all $a \in \mathcal{A}_+$ and all $\omega \in \mathcal{A}_+$*

$$\sup_{c,d \in \mathcal{E}} |\omega(c^*ad)| = \tilde{K}(\omega, a) = K(\omega, a) .$$

We now stick to our assumption $\omega \in \mathcal{A}_+$ but allow any $b \in \mathcal{A}$. Let $a = (b^*b)^{1/2}$. If the positive number s goes to zero, then $a^2(a + s\mathbf{1})^{-1}$ converges in norm to a. Hence $\omega(c^*ad)$ is the limit of $\omega(c_s^*bd)$ with $c_s = b(a + s\mathbf{1})^{-1} c$. But $c_s \in \mathcal{E}$ if $c \in \mathcal{E}$. Thus

$$\tilde{K}(\omega, a) \leqq \tilde{K}(\omega, b) .$$

Let now again $b, c \in \mathcal{E}$. It is $c^*bd = c^*b(a + s\mathbf{1})^{-1} (a + s\mathbf{1})\, d = c_s^*(a + s\mathbf{1})\, d$ and this time $c_s = (a + s\mathbf{1})^{-1}\, bc \in \mathcal{E}$ for positive s. Applying the definition of $\tilde{K}$ we find

$$\text{Real}\, \omega(c^*bd) \leqq K(\omega, a + s\mathbf{1}) .$$

But $s > 0$ is arbitrary and we get

$$\tilde{K}(\omega, b) \leqq \tilde{K}(\omega, a) .$$

Lemma 3-5. *Let ω be a positive linear functional and $a = (b^*b)^{1/2}$ with given $b \in \mathcal{A}$. Then*

$$\tilde{K}(\omega, b) = \tilde{K}(\omega, a) .$$

Corollary. *$\varrho \rhd \omega$ for two positive linear functionals if and only if $\tilde{K}(\varrho, a) \leqq \tilde{K}(\omega, a)$ for all $a \in \mathcal{A}_+$.*

We are allowed to replace the functionals $\tilde{K}$ by the Ky Fan functionals by virtue of Lemma 3-4. Comparing the result with the corollary of Theorem 3-2 we find

Theorem 3-6. *For two states $\varrho, \omega \in \mathcal{S}$ it is $\varrho \succ \omega$ if and only if $\varrho \rhd \omega$.*

3.3. Some essential results for W^*-algebras

Much more can be said about the partial order $\succ$ in the state space of W^*-algebras. Especially, one can derive for W^*-algebras a generalization of Theorem 2-2, i.e. non-commutative variants of the theorems of Rado and of Markus, see [13].

The key property for doing this will be stated below but will only be proved later in Chapter 5.

Theorem 3-7 ("Σ-property"). *Let $\mathcal{M}$ be a W^*-algebra and ω one of its positive linear functionals. For every finite set $s_1, \ldots, s_m$ of positive numbers and every increasing family $q_1 \leqq q_2 \leqq \cdots \leqq q_m$ of projection operators of $\mathcal{M}$ the equation*

$$\sum s_j K(\omega, q_j) = K(\omega, \sum s_j q_j)$$

is valid (see [6, 7, 11]).

The generality of this theorem is really amazing. The rather involved proof will be given in Chapter 5.

We shall draw some important consequences from Theorem 3-7. Let us assume $\varrho \rhd \omega$ for two positive linear functionals, ϱ, and ω. This means $K(\varrho, a) \leqq K(\omega, a)$ according to the corollary to Lemma 3-5, and to the Lemma 3-4, for all $a \in \mathcal{M}_+$. $\mathcal{M}$ being a W^*-algebra, the elements with finite spectrum in $\mathcal{M}_+$ constitute a norm-dense subset. Our family of inequalities between the corresponding Ky Fan functionals has therefore to be checked only for these elements. (The norm continuity of the Ky Fan functionals follows trivially from their definition.) If now we have

$$a = t_1 p_1 + \cdots + t_m p_m$$

with $t_1 \geqq t_2 \geqq \cdots \geqq t_m \geqq 0$ and mutually orthogonal projection operators, p_j, one can find non-negative numbers $s_1, \ldots, s_m$ by

$$\begin{aligned} t_1 &= s_1 + s_2 + \cdots + s_m\,, \\ t_2 &= \phantom{s_1 + {}} s_2 + \cdots + s_m\,, \\ &\ldots\ldots\ldots\ldots\ldots\ldots \\ t_j &= s_j + s_{j+1} + \cdots + s_m \end{aligned}$$

and construct projection operators

$$q_j = p_1 + p_2 + \cdots + p_j\,, \qquad j = 1, \ldots, m\,,$$

so that

$$a = s_1 q_1 + \cdots + s_m q_m$$

is a decomposition of the element a which satisfies the assumption of Theorem 3-7. It follows that $K(\varrho, a) \leqq K(\omega, a)$ is implied by $K(\varrho, q_j) \leqq K(\omega, q_j)$, $j = 1, 2, \ldots, m$. Hence we need only to check the inequalities with the Ky Fan functionals $K(., q)$, q being a projection operator! This gives the important Theorem 3-8. (See [2, 3, 5, 7, 10, 13, 97, 100, 104, 107].)

Theorem 3-8. *If $\varrho, \omega \in \mathcal{M}_+^*$ and if for every projection q of $\mathcal{M}$ the inequality $K(\varrho, q) \leqq K(\omega, q)$ is valid, then $\varrho \rhd \omega$. If, in addition, $\varrho(1) = \omega(1)$, then it is $\varrho \succ \omega$.*

4. The c-ideal

4.1. Definition and existence of the c-ideal

Throughout this part we adopt the notations introduced in Chapter 3, with the additional assumption that $\mathcal{M}$ be a properly infinite W^*-algebra with centre $\mathcal{Z}$. In $\mathcal{M}$ we concentrate attention on considering $*$-ideals intersecting the centre only trivially. Call the set of these ideals $C_{\mathcal{M}}$. Then, our aim is to assure us of the existence of $\mathcal{J}_\infty \in C_{\mathcal{M}}$ with property that $\mathcal{J} \subset \mathcal{J}_\infty$ whenever $\mathcal{J} \in C_{\mathcal{M}}$. In other words, we will show the existence of a uniquely determined maximal element in $C_{\mathcal{M}}$ if the latter is thought of as partially ordered by inclusion. Moreover, we shall see that $\mathcal{J}_\infty$ is uniformly closed.

It is this ideal $\mathcal{J}_\infty$ which will be named "c-ideal" in case of properly infinite $\mathcal{M}$ (see [8, 13]). The c-ideal will prove a powerful tool throughout all subsequently performed investigations. The proof on existence of $\mathcal{J}_\infty$ will be led by direct constructions. Besides, the proposed run of proof will provide us with a practicable characterization or "definition" of $\mathcal{J}_\infty$.

We start with some preliminary facts about uniformly closed ideals and the projection sets contained in it, in case of W^*-algebras. If $\mathcal{N}$ is a W^*-algebra, by $\mathcal{N}^p$ its projection lattice, with lattice operations $\vee$, $\wedge$ defined as usually, will be meant. By $\succsim$, $\sim$ v. Neumann's relation and the correspondingly equivalence in the projection lattice is denoted, and all definitions follow the standard ones. Because of the always clear context in usage of $\succsim$, we hope no confusion will arise from the double use of the symbol $\succsim$ (cf. 3.2.).

If $\mathcal{J}$ is an ideal in a W^*-algebra $\mathcal{N}$, by $\mathcal{J}^p$ we mean the set $\mathcal{J} \cap \mathcal{N}^p$. It is a known fact (cf. [19]) that any uniformly closed ideal $\mathcal{J}$ in a W^*-algebra $\mathcal{N}$ can be generated canonically by its projection set $\mathcal{J}^p$. This is due to the presence of the parallelogram law, i.e. $p \vee q - q \sim p - q \wedge p$ for all $p, q \in \mathcal{N}^p$, and since $r(a) \sim l(a)$ for all $a \in \mathcal{N}$, with $r(a)$ (resp. $l(a)$) the right (resp. left) support projection of a in $\mathcal{N}^p$. Furthermore, the set $\mathcal{J}^p$ of a $*$-ideal $\mathcal{J}$ in a W^*-algebra constitutes a so-called p-ideal, i.e.

$$\text{if } q \in \mathcal{J}^p \text{ and } q' \in \mathcal{N}^p \text{ with } q' \leq q, \text{ then } q' \in \mathcal{J}^p; \tag{1}$$

$$\text{if } q \in \mathcal{J}^p \text{ and } q' \in \mathcal{N}^p \text{ with } q' \sim q, \text{ then } q' \in \mathcal{J}^p; \tag{2}$$

$$\text{if } q, q' \in \mathcal{J}^p, \text{ then } q \vee q' \in \mathcal{J}^p \tag{3}$$

hold in $\mathcal{J}^p$. Moreover, there are mutually inverse bijections between the set of all p-ideals (i.e. projection sets obeying (1), (2), (3)) of $\mathcal{N}$ and the set of all uniformly closed ideals. The correspondences under discussion are defined by

$$\mathcal{J} \mapsto \mathcal{J}^p , \tag{4}$$

$$\mathcal{J}^p \mapsto \mathcal{J} = \text{uniform closure } \{a \in \mathcal{N}: l(a) \in \mathcal{J}^p\} . \tag{5}$$

By virtue of these facts, we will show that our discussions about elements of $C_{\mathcal{M}}$ may be shifted to considerations in the set of all p-ideals having only trivial intersection with the centre. Firstly, whenever $\mathcal{J}$ is a $*$-ideal in $\mathcal{M}$, with the help of a result the reader can find in the book [27] (Chapter III, § 5, 2., Lemma 9) we see that

$$(\text{uniform closure } \mathcal{J}) \cap \mathcal{Z} = \text{uniform closure } (\mathcal{J} \cap \mathcal{Z}) . \tag{6}$$

Thus, in taking the closure of elements in $C_{\mathcal{M}}$ we will not leave the set $C_{\mathcal{M}}$. Therefore, we are allowed to discuss in terms of closed ideals.

Let $\mathcal{P}$ be a p-ideal in $\mathcal{M}^p$. Then, by (4), (5) we find a uniformly closed ideal $\mathcal{J}$ in $\mathcal{M}$ such that $\mathcal{P} = \mathcal{J}^p$. Assume $\mathcal{P} \cap \mathcal{Z} = (o)$. Then, since $\mathcal{Z}$ is a W^*-algebra too, we have with the $*$-ideal $\mathcal{J}_0 = \{x \in \mathcal{M}: l(x) \in \mathcal{P}\}$ that $\mathcal{J}_0 \cap \mathcal{Z} = (o)$. Then, from (5) and (6) in application to $\mathcal{J}_0$ the relation $\mathcal{J} \cap \mathcal{Z} = (o)$ follows. On the other hand, from $\mathcal{J} \cap \mathcal{Z} = (o)$ for a $*$-ideal $\mathcal{J}$ the validity of $\mathcal{J}^p \cap \mathcal{Z} = (o)$ is obvious. Therefore, for closed ideal $\mathcal{J}$ it holds $\mathcal{J} \cap \mathcal{Z} = (o)$ if and only if $\mathcal{J}^p \cap \mathcal{Z} = (o)$.

The latter, together with the fact that (4) and (5) are bijections preservi nginclusion, make our problem of finding out $\mathcal{J}_\infty$ equivalent to an analogously formulated problem in terms of p-ideal which meet the centre only trivially. Hence, the following results are crucial (see [118]):

Lemma 4-1. *Let $\mathcal{M}$ be a properly infinite W^*-algebra. Then, for every p-ideal $\mathcal{J}^p$ with $\mathcal{J}^p \cap \mathcal{Z} = (o)$ we have*

$$\mathcal{J}^p \subset \{q \in \mathcal{M}^p: qz \not\sim z \text{ for all } z \in \mathcal{Z}^p, z \neq o\} .$$

Proof. We define $\mathcal{P} = \{q \in \mathcal{M}^p: qz \not\sim z \text{ for all } z \in \mathcal{Z}^p, z \neq o\}$, and show $\mathcal{P} \cap \mathcal{Z} = (o)$. Indeed, $\mathcal{P}$ is a set consisting of projections only which, by definition, excludes non-trivial central projections from being members of the set.

Let $\mathcal{J}^p$ be a p-ideal with $\mathcal{J}^p \cap \mathcal{Z} = (o)$. Assume $q \in \mathcal{J}^p$ with $qz \sim z$ for a $z \in \mathcal{Z}^p$. Then, from (1) comes that $qz \in \mathcal{J}^p$, and therefore $z \in \mathcal{J}^p$ by (2). The latter implies $z = o$. Hence $qz \not\sim z$ for any non-trivial central projection z, i.e. $q \in \mathcal{P}$, and $\mathcal{J}^p \subset \mathcal{P}$ is seen. ∎

Lemma 4-2. *Let $\mathcal{M}$ be a properly infinite W^*-algebra. Then,*

$$\mathcal{P} = \{q \in \mathcal{M}^p: qz \not\sim z \text{ for all } z \in \mathcal{Z}^p, z \neq o\}$$

is a p-ideal.

In the proof we will rely on a technical result (see [7, 8, 9]) which will prove to be very useful in other proofs, too.

Lemma 4-3. *Let $q, q' \in \mathcal{N}^p$ in a W^*-algebra $\mathcal{N}$. Assume at least one of q, q' to be properly infinite. Then, $q \succsim q'$ implies $q \vee q' \sim q$.*

Proof. Assume q' be properly infinite (i.e. for $z \in \mathcal{Z}^p$ with $zq' \neq o$ the projection zq' is infinite). Then there is a projection $q'' \leq q$, with $q'' \sim q'$, and mutually orthogonal projections p_1, p_2 exist such that $q'' = p_1 + p_2$ and $p_1 \sim p_2 \sim q''$.

Define $p = (q - q'') + p_1$. Then $pp_2 = o$ and $q = p + p_2$. Because of $q'' \sim p_1$ and $q = (q - q'') + q''$ we see $q \sim p$. From the parallelogram law $q \vee q' - q \sim q' - q \wedge q'$ comes that $q' \succsim q \vee q' - q$, so from $p_2 \sim q'' \sim q'$ the relation $p_2 \succsim q \vee q' - q$ can be seen. Finally, from $p \sim q$ and the derived facts we get $q = p_2 + p \succsim (q \vee q' - q) + q$, i.e. $q \succsim q \vee q'$ from which $q \sim q \vee q'$ follows.

Assume q is properly infinite. Then, there are mutually orthogonal projections p_1, p_2 fulfilling $q = p_1 + p_2$ and $p_1 \sim p_2 \sim q$. Again inserting the parallelogram law, we see $q \vee q' - q \sim q' - q \wedge q' \leqq q'$. Because of $q \sim p_2$, $q' \precsim q \sim p_1$ and mutual orthogonality of the projections under comparison we have to conclude that $q \vee q' = (q' \vee q - q) + q \precsim p_1 + p_2 = q$. The latter implies $q \sim q \vee q'$. ∎

We are going to show the validity of Lemma 4-2.

Proof of Lemma 4-2. By definition of $\mathcal{P}$, properties (1) and (2) for p-ideal are obviously present. The hard part is in verifying (3). Assume $q, q' \in \mathcal{P}$. We are going to show that in starting out from $q \vee q' \notin \mathcal{P}$ we will arrive at a contradiction.

Assume $q \vee q' \notin \mathcal{P}$. Then, by definition of $\mathcal{P}$, there is $z \in \mathcal{Z}^p$ such that

$$(q \vee q')\, z \sim z\,, \tag{7}$$

and $z \neq o$. Since z is a central projection, the triple of projections q, q', z is distributive in the projection lattice of $\mathcal{M}$, so the transcription of (7) into $qz \vee q'z \sim z$ is justified. $\mathcal{M}$ is properly infinite, so any central projection is infinite, too (if it is non-trivial). Hence, the projection $qz \vee q'z$ is infinite. This requires at least either qz or $q'z$ to be infinite (qz, $q'z$ both being finite would imply $qz \vee q'z$ to be finite in contradiction to (7)).

Assume qz is infinite. Then, there is a central projection z' such that $qzz' \neq o$ and qzz' is properly infinite (take for z' the central projection projecting on the properly infinite part of $q\mathcal{M}qz$). From (7) comes that

$$qzz' \vee q'zz' = (q \vee q')\, zz' \sim zz' \neq o\,. \tag{8}$$

The comparability theorem guarantees the existence of a central projection z'' with

$$z''z'zq \succsim z''z'zq'\,, \tag{9}$$

and

$$z''^{\perp}z'zq \precsim z''^{\perp}z'zq' \quad (\text{with } z'' = \mathbf{1} - z'')\,. \tag{10}$$

Assume $z''z'zq \neq o$. Since $z''z'zq$ is properly infinite, (9) makes that Lemma 4-3 becomes applicable with the result that

$$z''z'z(q \vee q') = z''z'zq \vee z''z'zq' \sim z''z'zq\,, \tag{11}$$

and in applying (7) to (11) we find

$$z''z'z \sim z''z'z(q \vee q') \sim z''z'zq\,. \tag{12}$$

Because of $z''z'z \neq o$ the result (12) violates the assumption that $q \in \mathcal{P}$.

Assume $z''z'zq = o$. Then, from $qzz' \neq o$ comes that $z''^{\perp}z'zq \neq o$. By (10) with $z''^{\perp}z'zq$ being properly infinite we are allowed to stress Lemma 4-3 once more (analogously to steps (11), (12)) with the result $z''^{\perp}z'z \sim z''^{\perp}z'zq' \neq o$. The latter contradicts $q' \in \mathcal{P}$. So we have arrived at contradictions under the assumption that qz be infinite. Since (7) is symmetric in q, q' the same line of proof applies under the assumption that $q'z$ is infinite, and contradictions will appear in this case, too. Therefore $q, q' \in \mathcal{P}$ has to imply $q \vee q' \in \mathcal{P}$, and (3) for p-ideal is justified in case of $\mathcal{P}$. By the remarks at the beginning of the proof we may then take Lemma 4-2 to be true. ▮

Taking the results of the preliminary discussions together with the statements of Lemma 4-1 and Lemma 4-2, we may formulate (see [118]):

Theorem 4-4. *Let $\mathcal{M}$ be a properly infinite W^*-algebra. There is a unique maximal element $\mathcal{J}_\infty$ in the set of $*$-ideals (of $\mathcal{M}$) having only trivial intersection with the centre $\mathcal{Z}$ of $\mathcal{M}$. $\mathcal{J}_\infty$ is uniformly closed and is generated by the p-ideal $\mathcal{J}_\infty^p = \{q \in \mathcal{M}^p\colon qz \not\sim z$ for all $z \in \mathcal{Z}^p, z \neq o\}$* i.e. $\mathcal{J}_\infty =$ *uniform closure* $\{x \in \mathcal{M}\colon l(x) \in \mathcal{J}_\infty^p\}$.

We call $\mathcal{J}_\infty$ c-ideal in the case of the properly infinite W^*-algebra $\mathcal{M}$. For convenience, we will adopt the following agreement:

> in the case of a general W^*-algebra $\mathcal{N}$ the c-ideal of the properly infinite part of $\mathcal{N}$ is referred to as the c-ideal of $\mathcal{N}$.

Thus, the c-ideal of a finite W^*-algebra is defined to be the trivial ideal.

We close this point with a technical result in context with c-ideals.

Lemma 4-5. *Let $\mathcal{M}$ be properly infinite. There is a uniquely determined central projection z, for $p \in \mathcal{M}^p$, such that $pz \sim z$ and $pz^\perp \in \mathcal{J}_\infty$.*

Proof. Collect the systems of mutually orthogonal central projections, with property that for any projection z of any system we have $zp \sim z$, into the set $\mathcal{F}_p$. Then, $\mathcal{F}_p$ is partially ordered by inclusion (increasing). Assume $(E_\lambda)_{\lambda \in \mathbf{I}} \subset \mathcal{F}_p$ to be linearily ordered ($\mathbf{I}$ — an index set) by inclusion. Take $E = \bigcup_{\lambda \in \mathbf{I}} E_\lambda$. Then, $E \in \mathcal{F}_p$ with $E \supset E_\lambda$ for all $\lambda \in \mathbf{I}$, so (since $(E_\lambda)_{\lambda \in \mathbf{I}}$ can represent any linearly ordered set) we may conclude that $\mathcal{F}_p$ is inductively ordered. Hence there is a maximal system E_∞ in $\mathcal{F}_p$. Assume $E_\infty = (z_i)_{i \in \mathbf{K}}$ with $\mathbf{K}$ being an index set. By definition of $\mathcal{F}_p$, $pz_i \sim z_i$ for all $i \in \mathbf{K}$, so $pz \sim z$ for $z = \sum_i z_i$. Maximallity of E_∞ then makes sure z to be maximal in the set of all central projections z' with $pz' \sim z'$, so $pz^\perp \in \mathcal{J}_\infty^p$, i.e. $pz^\perp \in \mathcal{J}_\infty$ is seen. Uniqueness is obvious. ▮

Notation (see [8, 118]). By the just derived results we are assured of a decomposition of a projection $p \in \mathcal{M}$ into mutually orthogonal projections pz, $pz^\perp$ with properties as above. This decomposition via the c-ideal in a properly infinite W^*-algebra will be referred to as *the canonical decomposition of p in $\mathcal{M}$.*

Remark. Let $\mathcal{M}$ be an infinite factor which is countably decomposable. Then, in case $\mathcal{M}$ is type III the c-ideal is trivial, and for a semi-finite $\mathcal{M}$ the respective c-ideal will be generated by the respective p-ideals of all finite projections.

4.2. Ky Fan functionals and von Neumann's relation

In this part, matter closely relates to notions introduced within 3.2., if seen in restriction to W^*-algebras and states. The aim is to establish a characterization of $\succ$, in the projection lattice of a general W^*-algebra $\mathcal{M}$, which reads in terms of the

Ky Fan functionals introduced in Chapter 3, (6). As will be proved later on in Chapter 5, the new aspect of looking on $\succsim$ in $\mathcal{M}^p$ via the Ky Fan functionals gives rise to an interesting duality principle between the two meanings of $\succsim$ we meet in context with W^*-algebras. On the one hand, we announced in 3.3. that $\omega \succsim \varrho$ for states ω, ϱ on a W^*-algebra $\mathcal{M}$ if and only if $K(\omega, p) \leqq K(\varrho, p)$ for all $p \in \mathcal{M}^p$; now, what about the relations between p and q, both being projections, in case of $K(\omega, p) \leqq K(\omega, q)$ for all $\omega \in \mathcal{S}$, with $\mathcal{S}$ the state space of $\mathcal{M}$? The answer is in [9, 13, 118].

Theorem 4-6. *In a W^*-algebra $\mathcal{M}$, for projections p, q we meet $p \succsim q$ iff $K(\omega, p) \geqq K(\omega, q)$ for all $\omega \in \mathcal{S}$.*

Proof. First, we comment on the necessity of $K(\omega, p) \geqq K(\omega, q)$ for all $\omega \in \mathcal{S}$ whenever $p \succsim q$.

Assume $p, q \in \mathcal{M}^p$, with $p \succsim q$. There is a projection q' with $q' \sim q$ and $p \geqq q'$. Since $K(\omega, a)$ depends monotonously on a with respect to $\leqq$ in $\mathcal{M}_h$, we see that $K(\omega, p) \geqq K(\omega, q')$ for every state ω. Therefore, if

$$q', q \in \mathcal{M}^p, \quad q' \sim q \quad \text{implies} \quad K(\omega, q') = K(\omega, q) \quad \text{for all } \omega \in \mathcal{S} \tag{13}$$

can be shown, this will do it. We derive (13). There is a partial isometry $w \in \mathcal{M}$ connecting q to q': $w^*w = q$, $ww^* = q'$, i.e. $w^*q'w = q$ and $wqw^* = q'$. Then, because of c^*w^*, $wd \in \mathcal{E}$ whenever $c, d \in \mathcal{E}$, with $\mathcal{E}$ denoting the unit sphere of $\mathcal{M}$, and due to Lemma 3-4, we see

$$\begin{aligned} K(\omega, q) &= \sup_{c,d \in \mathcal{E}} |\omega(c^*qd)| = \sup_{c,d \in \mathcal{E}} |\omega(c^*w^*q'wd)| \\ &\leqq \sup_{c',d' \in \mathcal{E}} |\omega(c'^*q'd')| = K(\omega, q') \quad \text{for all} \quad \omega \in \mathcal{S}. \end{aligned}$$

An analogous inequality is obtained if we start with q' instead of q, for the relation between q', q is a symmetric one. Thus, $K(\omega, q) \leqq K(\omega, q')$ and $K(\omega, q') \leqq K(\omega, q)$ for all $\omega \in \mathcal{S}$. Hence, (13) is true, and the necessity of the desired condition follows.

Next, we will prove the following statement, for projections $p, q \in \mathcal{M}^p$:

$$\text{if } p \succsim q, \text{ with } p \nsim q, \text{ then we find } \varrho \in \mathcal{S} \text{ with } K(\varrho, p) > K(\varrho, q). \tag{14}$$

The justification of (14) then gives the implication

$$K(\omega, p') \geqq K(\omega, q') \quad \text{for all } \omega \in \mathcal{S} \mapsto p' \succsim q' \quad \text{for } p', q' \in \mathcal{M}^p.$$

Indeed, accepting the validity of (14) for the moment, we might conclude as follows:

If $p', q' \in \mathcal{M}^p$ such that neither $p' \succsim q'$ nor $q' \succsim p'$ holds, the comparability theorem provides $z \in \mathcal{Z}^p$ with

$$p'z \succsim q'z, \quad p'z \nsim q'z; \qquad q'z^\perp \succsim p'z^\perp, \quad q'z^\perp \nsim p'z^\perp.$$

In applying (14) to the pairs of projections $p'z$, $q'z$ and $q'z^\perp$, $p'z^\perp$ respectively, we find states ϱ and ν with

$$K(\varrho, p'z) > K(\varrho, q'z) \quad \text{and} \quad K(\nu, q'z^\perp) > K(\nu, p'z^\perp) \tag{15}$$

respectively. Because of (15) and due to the construction of Ky Fan functionals,

$\varrho(z) \neq 0$, $\nu(z^\perp) \neq 0$ have to hold. Hence $\varrho' = \varrho(z)^{-1}\varrho(z.)$, $\nu' = \nu(z^\perp)^{-1}\nu(z^\perp.) \in \mathcal{S}$, and

$$K(\varrho', p') > K(\varrho', q'), \quad K(\nu', p') < K(\nu', q') \tag{16}$$

in respect to (15) once more. Therefore, comparability of p' with q' is necessary in order to avoid a case like (16), and the right direction of $\succcurlyeq$ between p', q' should be clear from principles (13) and (14).

Thus, all we have to demonstrate is the validity of (14). We remark that, due to (13), we are allowed to suppose $p > q$, $p \not\sim q$ in the assumption of (14). We adopt this point, and start with

First case: q is a finite projection.

Let $c(p)$ and $c(q)$ denote the central supports of p and q, respectively. Then, in case of $c(p) > c(q)$ the existence of a separating state ϱ with $K(\varrho, p) > K(\varrho, q)$ is trivial (take a state with support in $c(p)\,p - c(q)\,p$).

Let $c(p) = c(q)$. Then, $p - q$ and q cannot be disjunct to each other, i.e. we find a non-trivial projection $r \in \mathcal{M}^p$ with $r \leqq p - q$, $r \prec q$. Hence there is a finite $q'' \in \mathcal{M}^p$ with $p \geqq q'' > q$ (take $q'' = q + r$, for instance). The W^*-algebra $q''\mathcal{M}q''$ is finite, so there exists a tracial state τ on $q''\mathcal{M}q''$ with $\tau(q'' - q) \neq 0$. Define $\varrho = \tau(q''.q'') \in \mathcal{S}$. Due to $q''uq \in q''\mathcal{M}q''$ for every $u \in \mathcal{U}$, and since τ is tracial on $q''\mathcal{M}q''$, we may conclude that $\varrho(uqu^*) = \tau(q''uqu^*q'') = \tau(qu^*q''uq) \leqq \tau(q) = \varrho(q)$. The latter being valid for any unitary u implies $K(\varrho, q) = \varrho(q) = \tau(q) < \tau(q'') = 1$, by assumption on τ. On the other hand, $K(\varrho, p) \geqq K(\varrho, q'') \geqq \tau(q'') = 1$, so $K(\varrho, p) > K(\varrho, q)$ is seen.

Second case: p is properly infinite, $p > q$, $p \not\sim q$.

Then, $K(\omega, p) \geqq K(\omega, q)$ for all $\omega \in \mathcal{S}$ is clear. We show that the assumption $K(\omega, p) = K(\omega, q)$ for all $\omega \in \mathcal{S}$ raises a contradiction. In fact, the equality has to hold, particularly, in case of $\nu \in \mathcal{M}_+^*$ with structure $\nu = \omega(p.p)$, $\omega \in (p\mathcal{M}p)_+^*$: $K(\nu, p) = K(\nu, q)$, where we respected (8) of Chapter 3. Then, with $\mathcal{E}_p$ and $\mathcal{U}_p$ denoting the unit sphere and unitaries in $p\mathcal{M}p$ respectively, we see from Lemma 3-4 together with the definition of the Ky Fan functionals that

$$\begin{aligned}\nu(p) = \omega(p) = K\big(\omega(p.p), p\big) &= K\big(\omega(p.p), q\big)\\ &\leqq \sup_{u \in \mathcal{U}} \omega(pupqpu^*p) \leqq \sup_{a \in \mathcal{E}_p} \omega(aqa^*) = \sup_{a \in \mathcal{E}_p} \nu(aqa^*) \leqq \nu(p)\,,\end{aligned}$$

so equality has to occur. This, however, reads as

$$\omega(p) = \sup_{u \in \mathcal{U}_p} \omega^u(q) \quad \text{for all } \omega \in (p\mathcal{M}p)_+^*\,, \tag{17}$$

if Lemma 3-4 is applied in $p\mathcal{M}p$.

Let $q = qs^\perp + qs$ be the canonical decomposition of q with respect to the c-ideal $\mathcal{J}_p$ of $p\mathcal{M}p$, with $s \in \mathcal{Z}_p = \mathcal{Z}p$, and $qs \sim s$, $qs^\perp \in \mathcal{J}_p$ in $p\mathcal{M}p$.

Assume $\omega \in (p\mathcal{M}p)_+^*$, with $\omega(\mathcal{J}_p) = (0)$. Then, respecting the canonical decomposition of q, and applying (13) to the pair qs, s of projections, we see

$$\omega(s) = \sup_{u \in \mathcal{U}_p} \omega^u(q) \quad \text{for all } \omega \in (p\mathcal{M}p)_+^*\,, \quad \omega(\mathcal{J}_p) = (0)\,. \tag{18}$$

Taking together (17) with (18), we arrive at

$$\omega(p - s) = 0 \quad \text{for all } \omega \in (p\mathcal{M}p)_+\,, \quad \omega(\mathcal{J}_p) = (0) \tag{19}$$

In other terms, (19) reads as

$$\hat{\omega}(\widehat{p - s}) = 0 \quad \text{for all } \hat{\omega} \in (p\mathcal{M}p/\mathcal{J}_p)^*_+, \tag{20}$$

with $\widehat{p - s}$ denoting the equivalence class which $p - s$ is a representative of. Since $p\mathcal{M}p/\mathcal{J}_p$ is a C^*-algebra, by standard facts we know that (20) implies $\widehat{p - s} = \hat{o}$. By definition of the c-ideal, we have $\mathcal{J}_p \cap \mathcal{Z}_p = (o)$. This together with $p - s \in \mathcal{Z}_p$ proves $p - s$ to be the only central representative of $\widehat{p - s}$, so $\widehat{p - s} = \hat{o}$ gives $p - s = o$, i.e. $p = s$. The latter implies $q \sim p$ due to $qs \sim s$, $p > q$. So $q \sim p$ in $\mathcal{M}$, too. This violates one of the defining assumptions of this case. Therefore, a separating state has to exist in this situation, too.

With a view to principle (13), what we have derived up to now effectively is:

$$\text{for } p', q' \in \mathcal{M}^p \text{ obeying } p' \succ q', p' \nsim q', \text{ with } p' \text{ properly infinite or } q' \text{ finite, we find } \varrho \in \mathcal{S} \text{ such that } K(\varrho, p') > K(\varrho, q'). \tag{21}$$

In the third and last step, assume $p \succ q$, $p \nsim q$. To be non-trivial, assume $\mathcal{M}$ to be infinite. Let $z \in \mathcal{Z}^p$ be the projection on the properly infinite part of $\mathcal{M}$. Then, pz is properly infinite (or vanishes), and $pz^\perp$ is finite in any case. Our assumption $p \nsim q$ makes it that at least one of $pz \succ qz$ and $pz^\perp \succ qz^\perp$ is proper (i.e. excludes equivalence), so (21) applies in any situation in a non-trivial manner to one of the pairs pz, qz and $pz^\perp$, $qz^\perp$. Then, due to the structure of the Ky Fan functionals and $z, z^\perp \in \mathcal{Z}^p$, we find a non-trivial separating positive linear functional, so a separating state, for p, q as above, too. This completes the proof of (14). By our preliminary considerations we then can be sure of the validity of the implication $K(\omega, p) \geq K(\omega, q)$ for all $\omega \in \mathcal{S} \mapsto p \succ q$, in case of projections $p, q \in \mathcal{M}$. This closes the main proof. ∎

4.3. A centre-valued convex trace

In this part we will have to deal with a structure that proves to be very useful in context with order structure phenomena.

Definition 4-1. Let $\mathcal{M}$ be a W^*-algebra, with centre $\mathcal{Z}$, and assume T is a map from $\mathcal{M}_+$ into $\mathcal{Z}_+$. Then, the map T is referred to as a *centre-valued convex trace*, c.v.c.t. for short, if T has the following properties:

(i) $\|T(a)\| \leq \|a\|$ for all $a \in \mathcal{M}_+$;

(ii) $T(\mathbf{1}) = \mathbf{1}$;

(iii) $T(a + b) \leq T(a) + T(b)$ for all $a, b \in \mathcal{M}_+$;

(iv) $T(a) \geq T(b)$ for all $a \geq b \geq o$;

(v) $T(za) = zT(a)$ for all $a \in \mathcal{M}_+$, $z \in \mathcal{Z}_+$;

(vi) $T(a^*a) = T(aa^*)$ for all $a \in \mathcal{M}$.

As an example for a c.v.c.t. might serve the centre-valued trace in a finite W^*-algebra. We start with:

Lemma 4-7. *A centre-valued convex trace on a W^*-algebra is uniformly continuous.*

Proof. Let T be a c.v.c.t. on $\mathcal{M}$, and $a, b \in \mathcal{M}_+$. Then, $a + \|a - b\| \, 1 \geqq b \geqq o$, and by (iv), (iii), (ii) and homogeneity of T (being a trivial consequence of (v)) we see that

$$o \leqq T(a + \|a - b\| \, 1) - T(b) \leqq T(a) - T(b) + \|a - b\| \, 1 .$$

Hence $-\|a - b\| \, 1 \leqq T(a) - T(b)$. On the other hand, from the monotonicity of T together with (iii) and (ii) comes that

$$\begin{aligned} -\|a - b\| \, 1 \leqq T(a) - T(b) &\leqq T(a + \|a - b\| \, 1) - T(b) \\ &= T(b + (a - b) + \|a - b\| \, 1) - T(b) \\ &\leqq T(b) + T(a - b + \|a - b\| \, 1) - T(b) \\ &= T(a - b + \|a - b\| \, 1) \\ &\leqq \|a - b + \|a - b\| \, 1\| \, 1 \leqq 2 \, \|a - b\| \, 1 , \end{aligned}$$

so $\|T(a) - T(b)\| \leqq 2 \, \|a - b\|$, and uniform continuity is established. ∎

Besides (iii) and (iv), which are designed to serve as an appropriate substitute of "positive additivity", the Definition 4-1 exhibits the main properties we know from ♮ in a finite W^*-algebra. It is just the structure of this operation ♮ that provides an idea how to construct a peculiar c.v.c.t. Moreover, the construction scheme we will demonstrate applies in every W^*-algebra and in any case, results in a c.v.c.t. Therefore, we can be sure to find at least one c.v.c.t. in every W^*-algebra.

Let $\mathcal{M}$ be a W^*-algebra, with c-ideal $\mathcal{J}$ (remember our agreement on the meaning of c-ideal in general W^*-algebras, cf. below of Theorem 4-4). Then, $\hat{\mathcal{M}} = \mathcal{M}/\mathcal{J}$ is a C^*-algebra. Because of $\mathcal{Z} \cap \mathcal{J} = (o)$, $\mathcal{Z}$ may be thought of as isometrically embedded into the C^*-algebra $\hat{\mathcal{M}}$, and this embedding will be marked by $\hat{\mathcal{Z}}$. If $a \in \mathcal{M}$, by $\hat{a}$ we mean the element of $\hat{\mathcal{M}}$ generated by a via the ideal $\mathcal{J}$. In case of $\omega \in \mathcal{M}_+^*$, with $\omega(\mathcal{J}) = (0)$, by $\hat{\omega}$ we denote the corresponding element of $\hat{\mathcal{M}}_+^*$. We then know that every $\varrho \in \hat{\mathcal{M}}_+^*$ corresponds uniquely to a certain element of $\mathcal{M}_+^*$, with the same norm and vanishing on the c-ideal.

Whenever $a \in \mathcal{M}$, by $\mathbf{K}(\hat{a})$ we mean the uniformly closed convex hull of $C(\hat{a}) = \{b\hat{a}b^* : bb^* = 1, b \in \hat{\mathcal{M}}\}$ in $\hat{\mathcal{M}}$. Then, we have to take notice of the following result (cf. [118]):

Lemma 4-8. *To every $a \in \mathcal{M}_+$, there is a unique maximal element in $\hat{\mathcal{Z}} \cap \mathbf{K}(\hat{a})$. For the (uniquely) associated central element in $\mathcal{M}$ we write $a^\S$. Particularly, in case of a properly infinite $\mathcal{M}$, $a^\S$ satisfies the equation $\omega(a^\S) = K(\omega, a)$ for all $\omega \in \mathcal{M}_+$, $\omega(\mathcal{J}) = (0)$, and $\hat{a}^\S$ belongs to the uniform closure of $C(\hat{a})$.*

We postpone the proof to the end of this part, and are going to follow some consequences immediately.

Theorem 4-9. *In any W^*-algebra $\mathcal{M}$, the map $\S : \mathcal{M}_+ \ni a \mapsto a^\S$ is a centre-valued convex trace.*

Proof. Let z denote the central projection on the finite part of $\mathcal{M}$. Then, since the c-ideal has in common with $\mathcal{M}z$ only the zero element, we may write $\hat{\mathcal{M}} = \hat{\mathcal{M}}\hat{z}^\perp \oplus \mathcal{N}$, with $\mathcal{N}$ a finite W^*-algebra, which is isomorphic to $\mathcal{M}z$. So, there is a $\natural$-operation there, which will be denoted by the same symbol in both the cases. The $\S$-operation with respect to $\widehat{\mathcal{M}}\hat{z}^\perp$ will be marked by $\S'$.

Assume $a \in \mathcal{M}$. Since $\mathcal{M}z^\perp$ is properly infinite, by Lemma 4-8 we have

$$\widehat{az}^{\S'} = \|\text{-}\lim_n \hat{v}_n \hat{a} \hat{v}_n^* \quad \text{in } \widehat{\mathcal{M}}\hat{z}^\perp , \quad \text{with } \hat{v}_n \hat{v}_n^* = \hat{z}^\perp \text{ for all } n .$$

On the finite W^*-algebra $\mathcal{N}$, we know $(\widehat{az})^\natural$ to be the only central element in $\mathbf{K}(\widehat{az})$ (we notice that $\hat{v}\hat{z}$ is a unitary in $\mathcal{N}$ for $\hat{v}\hat{v}^* = \mathbf{1}$), so the $\S$-operation on $\mathcal{M}z$ has to equal $\natural$ on this algebra. Let $(\widehat{az})^\natural = \|\text{-}\lim_n \sum_{i \in \mathbf{I}} \lambda_n(i)\, \hat{u}_{in} \hat{a} \hat{u}_{in}^*$, with $\sum_{i \in \mathbf{I}} \lambda_n(i) = 1$, $\hat{u}_{in}\hat{u}_{in}^* = \hat{z}$, and $\mathbf{I}$ denoting an appropriate index set, and $\lambda_n(i) \geqq 0$ for all i and for every n, and at most finitely many indices of $\mathbf{I}$ with $\lambda_n(i) \neq 0$. We define $\hat{w}_{in} = \hat{v}_n + \hat{u}_{in}$. Then, $\hat{w}_{in}$ is a partial isometry in $\widehat{\mathcal{M}}$, with $\hat{w}_{in}\hat{w}_{in}^* = \mathbf{1}$ for all $i \in \mathbf{I}$ and for every n. With this set $\{\hat{w}_{in}\}$ we write $\widehat{(az^\perp)^{\S'} + (az)^\natural} = \|\text{-}\lim_n \sum_{i \in \mathbf{I}} \lambda_n(i)\, \hat{w}_{in} \hat{a} \hat{w}_{in}^*$ $\in \mathbf{K}(\hat{a})$, and have $(az^\perp)^{\S'} + (az)^\natural \in \mathcal{Z}$. But then, by construction of $\mathbf{K}(\hat{a})$, we must have that $a^\S = (az^\perp)^{\S'} + (az)^\natural$, $a^\S z^\perp = (az^\perp)^{\S'}$ and $a^\S z = (az)^\natural$, with $\S'$ and $\natural$ the $\S$-maps on the properly infinite and finite part of $\mathcal{M}$, respectively. The latter equalities mean that we may stop the proof after having shown the validity of the assertion in the two basic cases: finite and properly infinite W^*-algebras.

As we have learned, in the case of a finite W^*-algebra, $\S$ has to equal the operation of the centre-valued trace. This operation, however, is a c.v.c.t. trivially.

Suppose $\mathcal{M}$ is properly infinite. We have to check (i)—(vi) of Definition 4-1. By Lemma 4-8, we have $\omega(a^\S) = K(\omega, a)$ for all $\omega \in \mathcal{M}^*$, $\omega(\mathcal{J}) = (0)$. Then, by subadditivity, and homogeneity, and monotonicity of the Ky Fan functionals we get for a, b, c, d taken from $\mathcal{M}_+$, and $a \geqq b$, that

$$\omega(a^\S) \geqq \omega(b^\S) , \qquad \omega\big((c + d)^\S\big) \leqq \omega(c^\S) + \omega(d^\S)$$
$$\text{for all } \omega \in \mathcal{M}_+, \omega(\mathcal{J}) = (0) . \tag{21}$$

Because of $\mathcal{Z} \cap \mathcal{J} = (o)$, one has

$$\mathcal{Z}_+^* = \{\omega/_{\mathcal{Z}} : \omega \in \mathcal{M}_+, \omega(\mathcal{J}) = (0)\} , \tag{22}$$

so from (21) together with $a^\S, b^\S, c^\S, d^\S \in \mathcal{Z}_h$ the properties (iii) and (iv) follow, and presence of (ii) comes from the definition of the set $\mathbf{K}(\hat{a})$. To see (vi), let $b \in \mathcal{M}$, with polar decomposition $b = v/b/$, $/b/ = (b^*b)^{1/2}$. Then, $bb^* = v/b/^2 v^* = v(b^*b)\, v^*$, with $\|v\| = 1$. Hence $K(\omega, bb^*) \leqq K(\omega, b^*b)$ for all $\omega \in \mathcal{M}_+^*$, by Lemma 3-4. Since also $b^*b = w(bb^*)\, w^*$ with a partial isometry w, we see analogously $K(\omega, b^*b) \leqq K(\omega, bb^*)$. Therefore, $K(\omega, bb^*) = K(\omega, b^*b)$ for all $\omega \in \mathcal{M}_+^*$, $b \in \mathcal{M}$. Particularly the latter implies $\omega\big((bb^*)^\S\big) = \omega\big((b^*b)^\S\big)$ for all $\omega \in \mathcal{M}_+^*$, $\omega(\mathcal{J}) = (0)$ by Lemma 4-8. This means $(bb^*)^\S = (b^*b)^\S$, by (22).

To see (i), take a state ϱ on $\mathcal{Z}$, with $\varrho(a^\S) = \|a^\S\|$, $a \geqq o$. Then, there is a state ω on $\mathcal{M}$, with $\omega/_{\mathcal{Z}} = \varrho$, such that $\|a^\S\| = \varrho(a^\S) = \omega(a^\S) = K(\omega, a) \leqq \|a\|$ holds.

At last, we shall comment on property (v). We remark that (v) is valid for central projections. In fact, with arguments analogous to those we stressed at the beginning of the main proof, one becomes aware of

$$a^\S = (ac)^{\S_c} + (ac^\perp)^{\S_c\perp} , \qquad a^\S c = (ac)^{\S_c} , \qquad a^\S c^\perp = (ac^\perp)^{\S_c\perp} ,$$

with $c \in \mathscr{Z}^p$, and $\S_c$ the $\S$-operation on $\mathscr{M}c$, so $(ac)^\S = a^\S c$ can be followed.

Let $z \in \mathscr{Z}_+$, with $z = \sum_i \lambda_i c_i$, $\lambda_i \geqq 0$, $\{c_i\}$ mutually orthogonal projections in $\mathscr{Z}$. Then, for $\omega \in \mathscr{M}_+^*$, $\omega(\mathscr{J}) = (0)$,

$$\omega\big((za)^\S\big) = K(\omega, za) = \sum_i \lambda_i K(\omega, c_i a) = \sum_i \lambda_i \omega\big((c_i a)^\S\big)$$

$$= \sum_i \lambda_i \omega(c_i a^\S) = \omega(za^\S)\,, \qquad \text{i.e.} \quad (za)^\S = za^\S \qquad \text{for all } z$$

of a dense subset in $\mathscr{Z}_+$. Since (i)—(iv) now have been established, and we know $(\lambda a)^\S = \lambda a^\S$ for reals $\lambda \geqq 0$, we can be sure that $(.)^\S$ is uniformly continuous (remember that in the proof of Lemma 4-7 property (v) was inserted only insofar as $T(\lambda 1) = \lambda T(1)$, $\lambda \geqq 0$ was used!). But then, $za^\S = (za)^\S$ for all $z \in \mathscr{Z}_+$, too. Hence, (v) is established. In what we have proven, by the preliminary remarks, we can be sure of $\S$ being a c.v.c.t. also in the general case. ∎

Proposition 4-10. *On a W^*-algebra $\mathscr{M}$, with c-ideal $\mathscr{J}$, we have $(a^*a)^\S = o$ if and only if $a \in \mathscr{J}$.*

Proof. Due to the structure of $\S$, and since the centre-valued trace on a finite W^*-algebra is faithful, we may be content with treating the properly infinite case. Assume $\mathscr{M}$ is properly infinite, $\mathscr{J}$ the corresponding c-ideal.

Let $a \in \mathscr{J}$. Then, $a^*a \in \mathscr{J}$, so $\widehat{a^*a} = \hat{o}$. Hence $\mathbf{K}(\widehat{a^*a}) = \hat{o}$, i.e. $(a^*a)^\S = o$.

Let $b \in \mathscr{M}_+$, $b^\S = o$. Then $0 = \omega(b^\S) = K(\omega, b) \geqq \omega(b) \geqq 0$ for all $\omega \in \mathscr{M}_+^*$, $\omega(\mathscr{J}) = (0)$, i.e. $\omega(b) = 0$ there. The latter implies $\omega(b^{1/2}) = 0$ for all $\omega \in \mathscr{M}_+^*$, $\omega(\mathscr{J}) = (0)$, by the Schwarz-inequality. Hence, $\varrho(\widehat{b^{1/2}}) = 0$ for all $\varrho \in \hat{\mathscr{M}}_+^*$, so $\widehat{b^{1/2}} = \hat{o}$, i.e. $b^{1/2} \in \mathscr{J}$. Suppose $a^*a = b$. Then, what we derived is $|a| \in \mathscr{J}$, so $a = v\,|a| \in \mathscr{J}$ from the polar decomposition $v\,|a|$ of a. ∎

Proof of Lemma 4-8. We start with a properly infinite $\mathscr{M}$ and corresponding c-ideal $\mathscr{J}$. Assume $b \in \mathscr{M}_+$ such that

$$b = \sum \xi_i p_i\,, \tag{23}$$

with non-negative reals ξ_i and a finite ordered set of projections $o < p_1 < \cdots < p_n = 1$. Let $p_i = p_i z_i + p_i z_i^\perp$ mean the canonical decomposition of p_i, with

$$p_i z_i \sim z_i\,, \qquad p_i z_i^\perp \in \mathscr{J} \quad \text{for all } i\,. \tag{24}$$

Obviously, we have

$$o = z_0 \leqq z_1 \leqq \cdots \leqq z_n = 1\,. \tag{25}$$

Define $b_1 = \sum \xi_i p_i z_i$, $b_2 = \sum \xi_i p_i z_i$, $z = \sum \xi_i z_i$, and choose partial isometries u_k with $u_k u_k^* = z_k - z_{k-1}$, $u_k^* u_k = p_k z_k z_{k-1}^\perp$ $(k = 1, \ldots, n)$. Then, $u = \sum u_k$ is a partial isometry, with $uu^* = 1$. By an easy check one justifies from (24), (25) that

$$ub_1u^* = z\,, \quad \text{with} \quad uu^* = 1\,. \tag{26}$$

By construction, $b_2 \in \mathscr{J}$, so $ub_2u^* \in \mathscr{J}$, too. Hence, for all $\omega \in \mathscr{M}_+^*$, $\omega(\mathscr{J}) = (0)$:

$$\omega(z) = \omega(ubu^*) \leqq K(\omega, b)\,,$$

by Lemma 3-4. On the other hand, because of $p_i z \leqq z$ we have $ap_i z a^* \leqq aa^* z$, so $K(\omega, b_1) = \sup_{a \in \mathscr{E}} \omega(ab_1a^*) \leqq \omega(z)$. Since $ab_2a^* \in \mathscr{J}$, we then also have $K(\omega, b)$

$= K(\omega, b_1) \leqq \omega(z)$ in case of $\omega(\mathcal{J}) = (0)$. Taking together all that, we may conclude to

$$\omega(z) = K(\omega, b) \quad \text{for all } \omega \in \mathcal{M}_+^*, \omega(\mathcal{J}) = (0)\,, \tag{27}$$

with $\hat{z} \in C(\hat{b}) \cap \hat{\mathcal{Z}}$ by construction.

Assume $a \in \mathcal{M}_+$. Then with the spectral resolution $\{e(\lambda)\}$ of a, $a = \int_0^{\|a\|+\varepsilon} \lambda \, de(\lambda)$. Then, for instance, with

$$a_n = \|a\|(1 - e(\|a\|)) + \left(1 - \frac{1}{2^n}\right)\left\{e(\|a\|) - e\left(\left(1 - \frac{1}{2^n}\right)\|a\|\right)\right\} + \cdots$$
$$+ \frac{1}{2^n}\left\{e\left(\frac{1}{2^{n-1}}\|a\|\right) - e\left(\frac{1}{2^n}\|a\|\right)\right\},$$

we have

$$o \leqq a_1 \leqq a_2 \leqq \cdots \leqq a_n\,, \quad \text{uniform-}\lim_n a_n = a\,. \tag{28}$$

Then, every a_n can be transcribed such that (cf. 3.3.), for fixed n:

$$a_n = \sum \xi_i p_i\,, \qquad o < p_1 < \cdots < p_N = 1 \quad \text{projections,} \quad \xi_i \geqq 0\,, \tag{29}$$

i.e. a_n is "b-type" (see (23)). Fix another number m, with $m > n$. Then, due to the construction of the members of (a_k), we may write

$$a_m = \sum_i \sum_{j=1}^{N_i'} \xi_{ij} p_{ij}\,, \quad \text{with} \quad o < p_{11} < \cdots < p_{1N_1'} < \cdots < p_{NN_N'} = 1\,, \tag{30}$$

projections, and

$$p_{i1} = p_i\,, \qquad \sum_{j=1}^{N_i'} \xi_{ij} = \xi_i\,, \qquad \xi_{ij}\text{: non-negative reals.} \tag{31}$$

Proceeding with both a_n and a_m, as demonstrated in the case of b, we will arrive at $z(n)$ and $z(m)$, respectively, with (see (27))

$$\omega(z(n)) = K(\omega, a_n)\,, \quad \omega(z(m)) = K(\omega, a_m) \quad \text{for all } \omega \in \mathcal{M}^*, \ \omega(\mathcal{J}) = (0)\,. \tag{32}$$

Due to (29), (30) and (31) together with the construction of $z(n)$, $z(m)$, we have

$$z(m) \geqq z(n)\,, \qquad \|z(m) - z(n)\| \leqq \|a_m - a_n\|\,, \qquad m \geqq n\,. \tag{33}$$

We associate such a $z(k)$ to any a_k, and can be sure by (33) that $\{z(k)\}$ is a Cauchy sequence in $\mathcal{Z}_+$, so has to converge to some $z \in \mathcal{Z}_+$. The latter together with (28), (32) and the uniform continuity of Ky Fan functionals give

$$\omega(z) = K(\omega, a) \quad \text{for all } \omega \in \mathcal{M}, \ \omega(\mathcal{J}) = (0)\,. \tag{34}$$

By construction, there is $u_k \in \mathcal{M}$, with $u_k u_k^* = 1$, such that $z(k) = u_k a_k^1 u_k^*$, with a_k^1, a_k^2 corresponding to b_1, b_2 in case of b, respectively, so $u_k a_k^2 u_k^* \in \mathcal{J}$. This means, with $\|.\|^\wedge$ the norm in $\hat{\mathcal{M}}$, we have that

$$\begin{aligned} \|\hat{z} - \hat{u}_k \hat{a} \hat{u}_k^*\|^\wedge &\leqq \|\hat{z} - \hat{u}_k \hat{a}_k \hat{u}_k^*\|^\wedge + \|\hat{u}_k(\hat{a}_k - \hat{a})\,\hat{u}_k\|^\wedge \\ &\leqq \|\hat{z} - \hat{u}_k \hat{a}_k^1 \hat{u}_k - \hat{u}_k \hat{a}_k^2 \hat{u}_k\|^\wedge + \|\hat{a}_k - \hat{a}\|^\wedge \\ &\leqq \|\hat{z} - \widehat{z(k)}\|^\wedge + \|\hat{a}_k - \hat{a}\|^\wedge \\ &\leqq \|z - z(k)\| + \|a_k - a\| \underset{k\to\infty}{\rightarrow} 0\,, \end{aligned}$$

so we see the validity of

$$\hat{z} \in \text{uniform closure } C(\hat{a}) \cap \hat{\mathscr{Z}} \subset \mathbf{K}(\hat{a}) \cap \hat{\mathscr{Z}} . \tag{35}$$

Let $\hat{z}' \in \mathbf{K}(\hat{a}) \cap \hat{\mathscr{Z}}$. Then, $\hat{\omega}(\hat{z}') \leqq \hat{K}(\hat{\omega}, \hat{a})$ for all $\omega \in \mathscr{M}_+^*$, $\omega(\mathscr{J}) = (0)$, by construction of $\mathbf{K}(\hat{a})$ and structure of Ky Fan functionals ($\hat{K}$ means the Ky Fan functional in $\hat{\mathscr{M}}$). One easily proves (use Lemma 3-4 and Chapter 3, (6)) that $\hat{K}(\hat{\omega}, \hat{b}) = K(\omega, b)$ for all $b \in \mathscr{M}_+$, $\omega \in \mathscr{M}_+^*$, $\omega(\mathscr{J}) = (0)$, so we see from (34) that

$$\omega(z') = \hat{\omega}(\hat{z}') \leqq \hat{K}(\hat{\omega}, \hat{a}) = K(\omega, a) = \omega(z) \quad \text{for all } \omega \in \mathscr{M}_+^*,\ \omega(\mathscr{J}) = (0) .$$

This, however, by (22) implies $z' \leqq z$, so $\hat{z}' \leqq \hat{z}$, too. This proves $\hat{z}$ to be the unique maximal element of $\mathbf{K}(\hat{a}) \cap \hat{\mathscr{Z}}$. Finally, rename z into $a^{\S}$, and for properly infinite $\mathscr{M}$ the proof is closed.

For a finite W^*-algebra $\mathscr{M}$, we have $\mathscr{J} = (o)$ by definition, so $\hat{\mathscr{M}} = \mathscr{M}$, and $\mathbf{K}(\hat{a}) = \mathbf{K}(a)$ contains exactly one central element, namely $a^{\natural}$, where $\natural$ indicates the operation of taking the centre-valued trace. Thus, for every $a \in \mathscr{M}_h$, $a^{\natural}$ is, maximal in $\mathbf{K}(a) \cap \mathscr{Z}$, and $a^{\natural} = a^{\S}$ for $a \in \mathscr{M}_+$ has to be taken. The combination of both the cases we were dealing with then gives the existence of § with desired properties in the general case. In fact, define, for $a \in \mathscr{M}_+$, $a^{\S} = (az^{\perp})^{\S'} + (az)^{\natural}$, with §' the §-operation on the properly infinite part $\mathscr{M}z^{\perp}$, and $\natural$ the operation of centre-valued trace on the finite part $\mathscr{M}z$. In respecting $\widehat{(az^{\perp})^{\S'}} \in$ uniform closure $C(\hat{a}\hat{z}^{\perp}) \cap \hat{\mathscr{Z}}$, $(\widehat{az})^{\natural} \in \mathbf{K}(\widehat{az}) \cap \hat{\mathscr{Z}}$, one can easily follow that the so-defined $\hat{a}^{\S}$ is in $\mathbf{K}(\hat{a}) \cap \hat{\mathscr{Z}}$, and is the unique maximal element there (proceed in the details technically as in the first part of the proof of Theorem 4-9!). ∎

We concentrate the results of this part in a theorem:

Theorem 4-11. *On a W^*-algebra $\mathscr{M}$, with c-ideal $\mathscr{J}$, there is a map § from $\mathscr{M}_+$ into $\mathscr{Z}_+$ with*

(i) *§ is a centre-valued convex trace;*
(ii) *$(a^*a)^{\S} = o$ iff $a \in \mathscr{J}$;*
(iii) *in finite $\mathscr{M}$, § is the centre-valued trace;*
(iv) *in properly infinite $\mathscr{M}$, § is the maximal c.v.c.t. in the set*
$\{T\colon$ is c.v.c.t. with $T(a^*a) = o$ for all $a \in \mathscr{J}\}$.

Proof. We have only to comment on (iv). Let T be a c.v.c.t., with $T(a^*a) = o$ for all $a \in \mathscr{J}$. Take a projection p, with canonical decomposition $p = pz + pz^{\perp}$, $pz \sim z$, $pz^{\perp} \in \mathscr{J}$. Then $T(p) \leqq T(pz) + T(pz^{\perp}) = T(pz) = T(z) = zT(1) = z = p^{\S}$, for T is a c.v.c.t.. Take b as in (23). Then,

$$T(b) \leqq \sum \xi_i T(p_i) \leqq \sum \xi_i p_i^{\S} = b^{\S} ,$$

by facts known on §. Thus, $T(a) \leqq a^{\S}$ for elements a forming a uniformly dense subset of $\mathscr{M}_+$, so $T(h) \leqq h^{\S}$ for all $h \in \mathscr{M}_+$, by Lemma 4-7. ∎

4.4. Maximally mixed states (chaotic states)

Let $\mathscr{A}$ denote a C^*-algebra with $\mathbf{1}$, and the state space $\mathscr{S}$ considered with the relation $\succ$ as defined in Definition 3-1. We start with:

Definition 4-2. A state ω on $\mathcal{A}$ is said to be a *maximally mixed* or *chaotic* state, if ω is a member of the set $\mathcal{S}_\infty$ defined as

$$\mathcal{S}_\infty = \{\omega\colon \text{if } \varrho \in \mathcal{S} \text{ with } \varrho \succ \omega, \text{ so } \omega \succ \varrho\} .$$

We notify that $\mathcal{S}_\infty$ is non-void in any case. In fact, by definition of $\succ$ one easily recognizes $\{\mathcal{S}, \succ\}$ to be inductively ordered, so maximal states with respect to $\succ$ have to exist in any case. The latter, however, are exactly the members of $\mathcal{S}_\infty$. It is our task to clarify the structure of the set $\mathcal{S}_\infty$ in case of a W^*-algebra $\mathcal{M}$.

Theorem 4-12 (see [8], [118]). *On a W^*-algebra $\mathcal{M}$, we have $\omega(a^\S) \leqq K(\omega, a)$ for all $a \in \mathcal{M}_+$ and for all $\omega \in \mathcal{S}$. Moreover $\omega \in \mathcal{S}_\infty$ if and only if equality occurs, i.e.*

$$\omega(a^\S) = K(\omega, a) \quad \textit{for all } a \in \mathcal{M}_+ . \tag{36}$$

The latter means $\omega \in \mathcal{S}_\infty$ if and only if $\omega(\mathcal{J}) = (0)$, with $\mathcal{J}$ the c-ideal of $\mathcal{M}$, and the restriction of ω to the finite part of $\mathcal{M}$ is tracial.

Proof. Let $\mathcal{M}$ be properly infinite, and $b \in \mathcal{M}_+$ as in (23), with z, u, b_1, b_2 as defined in (25), (26), (27). Then, $z = b^\S$ and $ubu^* = b^\S + ub_2u^*$, with $b_2 \geqq o$. Since u is in the unit sphere, we have to follow for every state ω:

$$\omega(b^\S) \leqq \omega(ubu^*) \leqq K(\omega, b) .$$

The latter, with a view to uniform densedness of b-type elements in $\mathcal{M}_+$ and uniform continuity of § and Ky Fan functionals, implies

$$\omega(a^\S) \leqq K(\omega, a) \quad \text{for all } a \in \mathcal{M}_+ . \tag{37}$$

Let $\varrho \in \mathcal{S}$, with $\varrho(a^\S) = K(\varrho, a)$ for all $a \in \mathcal{M}_+$. Then, in the case of $\omega \succ \varrho$, we see from Lemma 3-4, Theorem 3-6 and (37) that

$$\varrho(a^\S) = K(\varrho, a) \geqq K(\omega, a) \geqq \omega(a^\S) .$$

Since $a^\S \in \mathcal{Z}$, and $\omega \in w^*$-closed convex hull $\{\varrho^v\colon v \in \mathcal{U}\}$, $\omega(a^\S) = \varrho(a^\S)$ has to hold, so $K(\omega, a) = K(\varrho, a)$ for all $a \in \mathcal{M}_+$. The latter means $\varrho \succ \omega$, i.e. $\varrho \sim \omega$, and $\varrho \in \mathcal{S}_\infty$.

On the other hand, by meaning of § and Lemma 4-8, we see for non-negative $a \in \mathcal{M}$ that $0 \leqq \varrho(a) \leqq K(\varrho, a) = \varrho(a^\S) = 0$ if $a \in \mathcal{J}$, so $\varrho(a) = 0$ for all $a \in \mathcal{J} \cap \mathcal{M}_+$, from which $\varrho(b) = 0$ for all $b \in \mathcal{J}$ follows. So, $\varrho(a^\S) = K(\varrho, a)$ implies $\varrho \in \mathcal{S}_\infty$ and $\varrho(\mathcal{J}) = (0)$. Since $\varrho(\mathcal{J}) = (0)$ implies $\varrho(a^\S) = K(\varrho, a)$ for all $a \in \mathcal{M}_+$, by Lemma 4-8, we are sure that

$$\varrho \in \mathcal{S}_\infty \quad \text{iff} \quad \varrho(\mathcal{J}) = (0) \quad \text{iff} \quad \varrho(a^\S) = K(\varrho, a) \quad \text{for all } a \in \mathcal{M}_+ ,$$

in case of a properly infinite $\mathcal{M}$.

Next, assume $\mathcal{M}$ is a finite W^*-algebra. Then, § equals the operation of the centre-valued trace on the positive cone, i.e.

$$a^\S = a^\natural \in \text{uniform closed convex hull } \{a^u\colon u \in \mathcal{U}\} ,$$

and the relation

$$\omega(a^\S) = \omega(a^\natural) \leqq K(\omega, a) \quad \text{for all } a \geqq o \tag{38}$$

is obvious, for every $\omega \in \mathcal{S}$.

Let $\varrho \in \mathcal{S}$, with $\varrho(a^\S) = \varrho(a^\natural) = K(\varrho, a)$. For $\omega \in \mathcal{S}$ with $\omega \succ \varrho$ from (38) we see $\varrho(a^\natural) = K(\varrho, a) \geqq K(\omega, a) = \omega(a^\natural)$, for, $\omega(a^\natural) = \varrho(a^\natural)$ due to $\omega \in w^*$-closure conv $\{\varrho^v\colon v \in \mathcal{U}\}$. So $\varrho \succ \omega$ is obvious (see Theorem 3-6), i.e. $\varrho \in \mathcal{S}_\infty$.

Let $\omega \in \mathcal{S}_\infty$. Look at $\omega^\natural = \omega((.)^\natural)$. Then, $K(\omega^\natural, a) = \omega^\natural(a^\natural) = \omega(a^\natural) \leqq K(\omega, a)$ for all $a \geqq o$, so $\omega \succsim \omega^\natural$. From this, however, $\omega = \omega^\natural$ can be followed with the definition of $\succsim$, if we refer to unitary invariance of $\omega^\natural$. Hence, $\mathcal{S}_\infty$ consists of tracial states exclusively. This proves the assertion in the finite case. Because of the structure of § as a c.v.c.t. and known facts on Ky Fan functionals, the general case then appears as a simple combination of the two cases we have been treating, and the assertion is true in any situation of a W^*-algebra. ∎

As a consequence of the just proven result we see (cf. [8]):

Corollary. *In every W^*-algebra, $\mathcal{S}_\infty$ is a w^*-closed convex subset of states, and $\mathcal{S}_\infty$ can be identified in a natural manner with the full state space of a particular C^*-algebra.*

It should be clear that the C^*-algebra in question is $\mathcal{A} = \mathcal{M}z/\mathcal{J} \oplus \mathcal{Z}z^\perp$, with z the central projection on the properly infinite part of $\mathcal{M}$. From this we easily see that $\mathcal{S}_\infty = \mathcal{S}$ may happen. The latter occurs iff either $\mathcal{J} = (o)$ or the algebra is commutative, or a central combination of both these cases in hand. The condition $\mathcal{J} = (o)$, however, in case that no commutative direct summand occurs, requires $\mathcal{M}$ to be type III and, in case we restrict considerations on von Neumann algebras on separable Hilbert spaces, $\mathcal{J} = (o)$ iff $\mathcal{M}$ is type III. This is due to the construction of the c-ideal and standard facts on type III algebras. We close this part by the remark that there is a significant difference between infinite and finite W^*-algebras, which manifests in the following:

> whereas in the finite case to every state ω corresponds exactly one chaotic state, namely $\omega^\natural$, in the infinite case a state admits many chaotic states, in general. If ω is the state in question, every $\varrho \in \mathcal{S}_\infty$ with $\varrho|_\mathcal{Z} = \omega|_\mathcal{Z}$ can be associated, and these states are numerous in general, and equivalent to each other in the sense of $\succsim$.

4.5. Technical Lemmata

In this part we derive some simple, technically minded, results in context with the notion of c-ideal and projections. The main field of application will be in the fifth chapter.

Lemma 4-13. *Let $\mathcal{M}$ be an infinite W^*-algebra. Assume p is a properly infinite projection, and q another projection, with $q < p$ and $q \in \mathcal{J}_p$, with $\mathcal{J}_p$ the c-ideal in $p\mathcal{M}p$. Then, we have $p - q \sim p$.*

Proof. By canonical decomposition of $p - q$ with respect to $\mathcal{J}_p$, and since $\mathcal{Z}p$ is the centre of $p\mathcal{M}p$, we find $z \in \mathcal{Z}^p$ such that $(p - q)\, z \sim pz$, and $(p - q)\, z^\perp \in \mathcal{J}_p$. Since $pz^\perp \in \mathcal{Z}p$ and $qz^\perp \in \mathcal{J}_p$ by assumption on q, we have $pz^\perp \in \mathcal{J}_p$, hence $pz^\perp = o$. This means $p \leqq z$, and from the above stated equivalence $p - q \sim p$ gives the result. ∎

Lemma 4-14. *Let $o = p_0 < p_1 < \cdots < p_n < p_{n+1} = 1$ and $o = p_0' < p_1' < \cdots < p_n' < p_{n+1}' = 1$ be properly infinite projections such that $p_i \in \mathcal{J}_{i+1}$ and $p_i \sim p_i'$*

for all i, with $\mathcal{J}_i$ denoting the c-ideal in $p_i\mathcal{M}p_i$. There is unitary $u \in \mathcal{U}$ with $p_i' = up_iu^*$ for all i.

Proof. Because of Lemma 4-13, we see $p_{i+1} - p_i \sim p_{i+1}$. Because $p_i \sim p_i$, we have: $p_i\mathcal{M}p_i$ is isomorphic to $p_i'\mathcal{M}p_i'$. Hence, $p_i' \in \mathcal{J}_{i+1}'$ and $p_{i+1}' - p_i' \sim p_{i+1}'$. This implies the equivalence $p_{i+1}' - p_i' = Q_i' \sim Q_i = p_{i+1} - p_i$ for all i. Define partial isometries v_i by $v_i^* v_i = Q_i$, $v_i v_i^* = Q_i'$. Then, since $\{Q_i\}$ and $\{Q_i'\}$ are orthogonal decompositions of $\mathbf{1}$, we recognize $u = \sum_i v_i$ to be unitary which fulfils all requirements. ∎

Lemma 4-15. *Let $\mathcal{M}$ be properly infinite, and $p_1 \leqq p_2 \leqq \cdots \leqq p_n \leqq \mathbf{1}$ an ordered set of properly infinite projections in $\mathcal{M}$. There is an orthogonal decomposition of $\mathbf{1}$ into projections $\{z_i\} \subset \mathcal{Z}^p$ such that either $p_k z_i \in \mathcal{J}_{p_{k+1}z_i}$ or $p_k z_i \sim p_{k+1} z_i$.*

Proof. In the first step, take $z \in \mathcal{Z}^p$ according to the canonical decomposition of p_n: $p_n z \sim z$, $p_n z^\perp \in \mathcal{J}_1 = \mathcal{J}$, and proceed with the ordered sets $p_1 z \leqq \cdots \leqq p_n z$ and $p_1 z^\perp \leqq \cdots \leqq p_n z^\perp$ in the properly infinite W^*-algebras $p_n\mathcal{M}p_n z$ and $p_n\mathcal{M}p_n z^\perp$ respectively (here, the trivial case is included) with respect to the associated c-ideals as in the first step we did for the unity, and so on. This defines an inductive process which — since we have only finitely many members — has to break up after performing at most 2^n steps, and the existence of the central decomposition of $\mathbf{1}$ associated to this process will do it.

Lemma 4-16. *Let $\mathcal{M}$ be properly infinite, and $p_1 \leqq \cdots \leqq p_n \leqq \mathbf{1} = p_{n+1}$, $p_1' \leqq \cdots \leqq p_n' \leqq \mathbf{1} = p_{n+1}'$ ordered sets of properly infinite projections, with $p_i \sim p_i'$. There is a partial isometry $v \in \mathcal{M}$ with $vv^* = \mathbf{1}$ and $vp_iv^* \geqq p_i'$ for all i, $v^*v \geqq p_1$.*

Proof. Let $\{z_j\}$ denote the central decomposition constructed along the line given in Lemma 4-15. Fix j for the moment, and look at $p_1 z_j \leqq \cdots \leqq p_n z_j \leqq z_j \leqq z_j p_{n+1}$. We choose indices $k_1 < \cdots < k_l$ such that $k_r \in \{1, \ldots, n+1\}$, with $p_{k_r} z_j \neq o$, and the following two conditions are fulfilled:

$$p_s z_j \in \mathcal{J}_{p_{k_r} z_j} \text{ for } s < k_r, \text{ and } p_s z_j \sim p_{k_r} z_j \text{ for } k_r \leqq s < k_{r+1}$$
$$\text{in case of } r \leqq l - 1 \text{ or } s \geqq k_l \ (r = l). \tag{39}$$

Obviously, the number l depends on the index j. Moreover, we require that

$$k_1 = \min_{p_s z_j \neq o} s\,. \tag{40}$$

Because of the relation $p_{k_l} z_j \sim p_{k_l}' z_j \sim p_{k+1}' z_j = z_j$ we find a partial isometry w with

$$w^*w = p_{k_l} z_j\,, \qquad p_{n+1}' z_j = z_j = ww^*\,. \tag{41}$$

Look at the projections defined as $p_s'' = wp_sw^*$, and compare the ordered sets of projections $p_{k_1}'' < \cdots < p_{k_l}'' = p_{n+1}' z_j = z_j$ and $p_{k_2-1}' z_j < \cdots < p_{k_l-1}' z_j < p_{n+1}' z_j = z_j$. Then, the assumptions and choice of the k_r's show that $p_{k_r}'' \sim p_{k_{r+1}}' z_j$ for all r (see (39)), and the two ordered sets of projections obey conditions for an application of Lemma 4-14 with respect to $\mathcal{M}z_j$. Hence, there is $u \in \mathcal{M}z_j$ with

$$u^*u = uu^* = z_j\,, \qquad up_{k_r}''u^* = p_{k_{r+1}-1}' z_j\,, \qquad up_{k_l}''u^* = z_j\,.$$

We define $u_j = uw$, and see the validity of

$$u_j^* u_j = p_{k_l} z_j\,, \qquad u_j u_j^* = z_j\,, \qquad u_j p_{k_r} u_j^* = p_{k_{r+1}-1}' z_j\,, \qquad u_j p_{k_l} u_j^* = z_j\,.$$

The latter gives, under the assumption $k_r \leqq s < k_{r+1}$ (or $s \geqq k_l$),

$$z_j p_s' \leqq z_j p_{k_{r+1}}' = u_j p_{k_r} u_j^* \leqq u_j p_s u_j^* \leqq z_j \,, \tag{42}$$

where $p_{k_r} \leqq p_s$ and $p_s' \leqq p_{k_{r+1}-1}'$ have been respected.

Because of (40), however, (42) is valid for every $s \in \{1, 2, \ldots, n+1\}$ (indeed, $p_s z_j \neq o$ requires $s \geqq k_1$).

We can construct an element $u_j \in \mathcal{M} z_j$, with (42) fulfilled, for every j. These add up to a partial isometry v, with

$$p_s' = v p_s v^* \,, \qquad v v^* = 1 \,, \qquad v^* v \geqq p_1 \,,$$

where on has to refer to (42) and the construction of u_j. ∎

5. The Σ-property

5.1. The definition, preliminary remarks

Let $\mathcal{M}$ be a W^*-algebra. For a positive linear form ω on $\mathcal{M}$, let there be defined (see [4, 10, 13]):

Definition 5-1. ω has the Σ-property if the additivity law $\sum_j s_j K(\omega, q_j) = K(\omega, \Sigma s_j q_j)$ holds for every finite set $\{q_j\}$ of projections, with $q_1 \leqq \cdots \leqq q_m$, and any choice of reals $s_1, \ldots, s_m \geqq 0$.

A moments reflection shows that ω has Σ-property if and only if

$$K(\omega, \Sigma q_j) = \Sigma K(\omega, q_j) \tag{1}$$

for every finite, linearly ordered family $\{q_j\}$ of projections. The main objective of this part is to show that Σ-property is intrinsic, i.e. it will prove to be a universal property (see [7]):

Theorem 5-1. *Every positive linear functional on a W^*-algebra has the Σ-property.*

The decisive step towards the justification of the formulated result will be in the following:

Lemma 5-2. *Every normal positive linear form on a W^*-algebra has the Σ-property.*

The proof of Lemma 5-2 will be performed throughout Sections 2—4.

The importance of Lemma 5-2 is due to the observation that Lemma 5-2 implies the statement of Theorem 5-1 to be true, i.e.

$$\text{Lemma 5-2} \rightarrow \text{Theorem 5-1}\,. \tag{2}$$

In this section, it will be our task to demonstrate the validity of implication (2). In order to do this, and for further use, we are going to fix some additional notations. Let $\mathcal{M}$ be a W^*-algebra. Then, we may think of $\mathcal{M}$ as canonically embedded into its second dual $\mathcal{M}^{**}$, which is a concrete W^*-algebra, and

$$\mathcal{M}^{**} = \text{strong closure } \mathcal{M}$$

if $\mathcal{M}$ is understood in its canonical realization on the $\mathcal{M}^{**}$ underlying Hilbert space. In this context, every state of $\mathcal{M}$ can be thought of as the restriction of a uniquely determined normal state of $\mathcal{M}^{**}$, i.e. $\mathcal{M}^*$ can be identified with the predual $(\mathcal{M}^{**})_*$ of $\mathcal{M}^{**}$ in the usual canonical manner. If $\omega \in \mathcal{M}^*$, the same letter ω will be used to denote the corresponding normal functional on $\mathcal{M}^{**}$. Let $\mathcal{U}^{**}$ and $\mathcal{E}^{**}$ stand for the unitaries and the unit sphere, respectively, in the W^*-

algebra $\mathcal{M}^{**}$. Then, we may define the Ky Fan functional K^{**} on $\mathcal{M}^{**}$ as in Chapter 3, (6), and have as a consequence of Lemma 3-4 and Chapter 3, (6):

$$K^{**}(\varrho, a) = \sup_{u \in \mathcal{U}^{**}} \varrho^u(a) = \sup_{b \in \mathcal{E}^{**}} \varrho^b(a) \tag{3}$$

for $\varrho \in (\mathcal{M}^{**})^*_+$, $a \in \mathcal{M}^{**}_+$.

Now, let us adopt the following agreement: whenever $a \in \mathcal{M}^{**}_+$ and $\varrho \in (\mathcal{M}^{**})^*_+$, we put

$$K(\varrho, a) = \sup_{u \in \mathcal{U}} \varrho^u(a)\ , \tag{4}$$

with the unitary group $\mathcal{U}$ of $\mathcal{M}$, which reduces to the Ky Fan functional on $\mathcal{M}$ in case of $\varrho \in \mathcal{M}^*_+$, $a \in \mathcal{M}_+$.

In remembering the way we arrived at Lemma 3-4, one becomes aware of

$$K(\varrho, a) = \sup_{u \in \mathcal{U}} \varrho^u(a) = \sup_{b \in \mathcal{E}} \varrho^b(a) \tag{5}$$

for *all* $\varrho \in (\mathcal{M}^{**})^*_+$, $a \in \mathcal{M}^{**}_+$, in this more general context, too.

Now, assume $\omega \in \mathcal{M}^*_+$. Then, in respecting Kaplansky's density theorem, we have $\mathcal{E}^{**} =$ strong closure $\mathcal{E}$, so we might conclude as follows: for $a \in \mathcal{M}^{**}_+$, $K^{**}(\omega, a) = \sup_{b \in \mathcal{E}^{**}} \omega^b(a) = \sup_{b \in \mathcal{E}} \omega^b(a) = K(\omega, a)$, i.e.

$$K(\omega, a) = K^{**}(\omega, a) \quad \text{for all} \quad \omega \in \mathcal{M}^*_+\ , \quad a \in \mathcal{M}^{**}_+\ , \tag{6}$$

with K, K^{**} defined as in (4), (3). Especially, in restriction to $\mathcal{M}$ the latter reads as

$$K(\omega, a) = K^{**}(\omega, a) \qquad \text{for all} \quad \omega \in \mathcal{M}^*_+\ , \quad a \in \mathcal{M}_+\ , \tag{7}$$

connecting the Ky Fan functionals of the two *different* algebras, with the canonical interpretation adopted on the right-hand side of (7). The proof of (2) will then be a first application of (7). Let $q_1 \leqq \cdots \leqq q_m$ be projections of $\mathcal{M}$. Then, $\{q_j\} \subset \mathcal{M}^{**p}$, too. Since $\omega \in \mathcal{M}^*_+$ is *normal* on $\mathcal{M}^{**}$, and Σ-property is granted by Lemma 5-2 to every normal positive linear form on a W^*-algebra, we have $K^{**}(\omega, \sum q_j) = \sum K^{**}(\omega, q_j)$. The latter, however, reduces to $K(\omega, \sum q_j) = \sum K(\omega, q_j)$ in virtue of (7). This holds for any choice of $q_1 \leqq \cdots \leqq q_m$ from the projection lattice of $\mathcal{M}$, so ω has Σ-property. This proves implication (2) to be true.

We showed that Theorem 5-1 is true under the assumption that Lemma 5-2 is valid. The proof of the latter, however, has to be accomplished without any reference to particular (and usually very useful) additional assumptions with respect to decomposition properties. Therefore, the tools we have to favour for a proof of Lemma 5-2 can only be a few very "classical" ones which refer to the projection lattice, such as the comparability theorem and basic facts on isometries.

5.2. The case of properly infinite projections — type III

Let $\mathcal{M}$ be a properly infinite W^*-algebra. Assume $p_1 < \cdots < p_n < p_{n+1} = \mathbf{1}$ is a finite set of properly infinite projections. Suppose $\omega \in \mathcal{M}^*_+$ and fix a real $\varepsilon > 0$. Then, we find projections $\{p_i'\}$, with $p_{n+1}' = \mathbf{1}$, such that

$$K(\omega, p_i) \geqq \omega(p_i') \geqq K(\omega, p_i) - \varepsilon\ , \qquad p_i' \sim p_i \quad \text{for all} \quad i\ , \tag{8}$$

where Chapter 4, (13), has been respected. In general, $\{p_i'\}$ is no longer linearily ordered by $\leqq$. Define projections $q_k = \bigvee_{i=1}^{k} p_i'$ for all k. Then, $q_1 \leqq \cdots \leqq q_n \leqq q_{n+1} = 1$. Because of $p_1' \precsim p_2' \precsim \cdots \precsim p_n'$ and since all these projections are properly infinite ones, we are allowed to apply Lemma 4-3 repeatedly, with the result that $q_k \sim p_k' \sim p_k$ for all k. The latter, together with (8) and Chapter 4, (13), shows the validity of

$$K(\omega, p_i) \geqq \omega(q_i) \geqq K(\omega, p_i) - \varepsilon \quad \text{for all} \quad i \leqq n \,. \tag{9}$$

In applying Lemma 4-16 to the ordered sets $p_1 < \cdots < p_{n+1} = 1$, $q_1 \leqq \cdots \leqq q_{n+1} = 1$, we are assured of the existence of a partial isometry v, with $vp_iv^* \geqq q_i$ for all i. Hence, by (9), Chapter 4, (13), and the fact that v is in $\mathscr{E}$ (and Lemma 3-4 holds), we may follow that

$$\sum_i K(\omega, p_i) \geqq \omega\left(v\left(\sum_i p_i\right)v^*\right) \geqq \sum_i K(\omega, p_i) - n\varepsilon \,. \tag{10}$$

On the other hand, due to $v \in \mathscr{E}$ and subadditivity of Ky Fan functionals, we have

$$\sum_i K(\omega, p_i) \geqq K\left(\omega, \sum_i p_i\right) \geqq \omega\left(v\left(\sum_i p_i\right)v^*\right).$$

This implies by (10):

$$0 \leqq \sum_i K(\omega, p_i) - K\left(\omega, \sum_i p_i\right) \leqq n\varepsilon \,.$$

Since n is constant, and $\varepsilon > 0$ could be chosen as small as we like, $\sum_i K(\omega, p_i) = K\left(\omega, \sum_i p_i\right)$ has to hold. The latter must be required for every finit eordered set of projections and arbitrarily chosen $\omega \in \mathscr{M}_+^*$.

Now, if $p_1 < \cdots < p_n$ are properly infinite projections in a general W^*-algebra, they have to be in the properly infinite part of the algebra in question, and the just realized result also proves the following result to be right:

Lemma 5-3 (see [7, 97]). *Let $\mathscr{M}$ be an infinite W^*-algebra, and $\{p_i\}$ a finite ordered set of properly infinite projections. Then, for every $\omega \in \mathscr{M}_+^*$ we have $K\left(\omega, \sum_i p_i\right) = \sum_i K(\omega, p_i)$. Especially, in case of purely infinite $\mathscr{M}$ this proves Theorem* 5-1.

Two things have to be noticed: firstly, in the case at hand, we did not have to make any reference to normality; secondly, the crucial structures, which have been inserted and allowed us to handle so freely through all purely infinite cases, are the c-ideal and the canonical decomposition which are the main ingredients of Lemma 4-16.

5.3 The finite case

In this section we concentrate on proving Lemma 5-2 in case of a finite W^*-algebra. Let $\mathscr{M}$ be a finite W^*-algebra. Then, by standard theory, there is a faithful family of normal, tracial states on $\mathscr{M}$. A normal, tracial state has its support in the centre. Look at the set of families of normal, tracial states, the supports of which are mutually orthogonal central projections. We may think of the set in question as partially ordered by inclusion (increasingly). Then, the set is an inductively ordered one. So there exists a maximal family $(\tau_i)_{i \in \mathbf{I}}$ in this set ($\mathbf{I}$ is an appropriate index set). Maximality implies $(\tau_i)_{i \in \mathbf{I}}$ to be faithful. Denote the set of all finite subsets of $\mathbf{I}$ by $\mathbf{P}$, and associate with $J \in \mathbf{P}$ a tracial, normal functional $t_J = \sum_{i \in J} \tau_i$. Then, the set $\mathcal{N} = \{t_J(a.)\}_{a \in \mathcal{M}, J \in \mathbf{P}}$ is a linear subspace of the predual $\mathcal{M}_*$. By construction of $\mathcal{N}$ it separates the points of $\mathcal{M}$. Therefore, due to the uniqueness of the predual, we must have that

$$\mathcal{M}_* = \text{uniform closure } \mathcal{N} \quad \text{(in functional norm)}\,.$$

Hence, with a view to uniform continuity of Ky Fan functionals, we can be sure that the Σ-property for elements of $\mathcal{M}_{*+}$ can be followed if it can be proved for elements of $\mathcal{N}_+$, i.e. for functionals of type $\tau(a.)$, with τ a normal, tracial state and $a \in \mathcal{M}_+$. Once more because of continuity, we may restrict considerations on such a which are of the form $a = \sum_j s_j p_j$, with $p_1 < \cdots < p_n$ projections, and reals $s_1, \ldots, s_n \geqq 0$. Since every normal state on the centre of $\mathcal{M}$ may be extended to a normal tracial state on $\mathcal{M}$, the required proof is a direct consequence of

Lemma 5-4. *Let $q_1 < \cdots < q_m$, $p_1 < \cdots < p_n$ be finite sets of projections in the finite W^*-algebra $\mathcal{M}$, with operation of the centre-valued trace $\natural$. Then, there is a unitary $u \in \mathcal{U}$ such that, for any choice of reals $s_1', \ldots, s_m', s_1, \ldots, s_n \geqq 0$, with $b = \sum_i s_i' q_i$ and $a = \sum_j s_j p_j$ we have that*

$$(aubu^*)^\natural \geqq (avbv^*)^\natural \quad \textit{for all} \quad v \in \mathcal{U}\,.$$

Proof. Taking the Comparability theorem (which works for two projections) as the first step of an inductive process, one easily succeeds in extracting a finite, orthogonal decomposition (z_k) of the identity into central projections z_k such that the members of $\{p_r z_k, q_s z_k\}_{r,s}$ are mutually comparable projections for each index k. Once more arguing inductively, we provide a unitary u_k such that the set $\{p_r z_k, u_k q_s u_k z_k\}_{r,s}$ becomes linearily ordered (in sense of $\leqq$). We omit the details, for, in a finite W^*-algebra, any equivalence can be realized by a unitary element.

Now, from properties of $\natural$ it follows that

$$(p_r v q_s v^*)^\natural z_k \leqq p_r^\natural z_k\,, \qquad (p_r v q_s v^*)^\natural z_k \leqq q_s^\natural z_k \quad \text{for all} \quad v \in \mathcal{U}$$
$$\text{and for all } r, s\,. \tag{11}$$

Fix r, s. Since $p_r z_k$, $q_s z_k$ are comparable to each other, we may assume $p_r z_k \precsim q_s z_k$. Then, due to the choice of u_k we have $p_r z_k \leqq u_k q_s u_k^* z_k$, so $(p_r u_k q_s u_k^*)^\natural z_k = p_r^\natural z_k$. The latter and (11) give, due to $p_r^\natural z_k \leqq q_s^\natural z_k$, the validity of

$$(p_r v q_s v^*)^\natural z_k \leqq (p_r u_k q_s u_k^*)^\natural z_k\,. \tag{12}$$

In case of $p_r z_k \succsim q_s z_k$ the argumentation is similar and again leads to (12).

We define $u = \sum_k u_k z_k$. Then, $u \in \mathcal{U}$ and from (12) comes that $(p_r v q_s v^*)^\natural \leqq (p_r u q_s u^*)^\natural$ for all $v \in \mathcal{U}$, where we had to take into consideration that (z_k) forms a decomposition of the identity. Finally, because of $s_s', s_r \geqq 0$ we see $(avbv^*)^\natural = \sum_{r,s} s_r s_s' (p_r v q_s v^*)^\natural \leqq \sum_{r,s} s_r s_s' (p_r u q_s u^*)^\natural = (aubu^*)^\natural$ for all $v \in \mathcal{U}$. ∎

Since the unitary in Lemma 5-4 does not depend on the special choice of the s_j''s, we can be assured that every $\tau(a.) = \omega$ has the property $K\left(\omega, \sum_j q_j\right) = \sum_j K(\omega, q_j)$. So, by our preliminary remarks, we may take for proof:

Lemma 5-5. *Every normal, positive linear functional on a finite W^*-algebra has Σ-property.*

5.4. The universality of the Σ-property for W^*-algebras

In this section we close the proof of Lemma 5-2, and so the proof of Theorem 5-1.

Lemma 5-6. *Let $\mathcal{M}$ be a properly infinite and semi-finite W^*-algebra. Every positive linear form on $\mathcal{M}$, which is normal, possesses the Σ-property.*

Proof. At first we remark that it will be sufficient to handle with sequences of projections, the members of which are finite or properly infinite. Indeed, let $Q_1 < \cdots < Q_k$ be projections in $\mathcal{M}$. Then there exist central projections $z_1, \ldots, z_k$ such that $Q_i\mathcal{M}Q_iz_i$ is properly infinite or trivial, and $Q_i\mathcal{M}Q_iz_i^\perp$ is finite, so Q_iz_i is properly infinite or trivial, and $Q_iz_i^\perp$ is finite. Obviously $z_1 \leqq z_2 \leqq \cdots \leqq z_k$. Setting $c_1 = z_1$, $c_i = z_i - z_{i-1}$ for $i = 2, \ldots, k$, and $c_{k+1} = z_k^\perp$, and considering $Q_1c_i \leqq \cdots \leqq Q_kc_i$ we notice that Q_jc_i is finite in case of $j \leqq i - 1$ (with $i = 2, 3, \ldots, k, k+1$), and is properly infinite or trivial for $j \geqq i$ (with $i = 1, \ldots, k+1$). The family $\{c_j\}$ of central projections forms a decomposition of the identity, so we may conclude that $K(\omega, Q_j) = \sum_i K(\omega, Q_jc_i)$, and the additivity law is valid if it proves to be true for the $\{Q_jc_i\}$-sequences of projections, with the property we notified.

In line with this, let $q_1 < \cdots < q_m$ be finite projections of $\mathcal{M}$, and assume $p_1 < \cdots < p_n$ to be properly infinite, with $q_m < p_1$. Assume ω to be a normal, positive linear functional of $\mathcal{M}$. Fix a real $\varepsilon > 0$.

By Lemma 5-3 we are provided with $u \in \mathcal{U}$ such that

$$\omega\left(u\left(\sum_i p_i\right) u^*\right) \geqq \sum_i K(\omega, p_i) - \varepsilon\,. \tag{13}$$

Define projections by

$$p_i' = up_iu^*\,, \qquad q_j' = uq_ju^* \quad \text{for all} \quad i, j\,. \tag{14}$$

Let $\{q_j''\}$ be a sequence of projections, with $q_j'' \sim q_j \sim q_j'$, and

$$\omega(q_j'') \geqq K(\omega, q_j) - \varepsilon\,. \tag{15}$$

The existence of q_j'' is clear by definition of Ky Fan functionals. We construct a finite projection q by $q = \bigvee_{j=1}^{m} q_j''$, and define

$$P_i = p_i' \vee q \quad \text{for all } i\,, \quad \text{and have} \quad P_i \sim p_i' \sim p_i\,. \tag{16}$$

Indeed, because of $q_1'' \precsim q_2'' \precsim \cdots \precsim q_m'' \precsim p_1' \leqq \cdots \leqq p_n'$ we see from an application of Lemma 4-3 that $p_i' \vee q_1'' \sim p_i' \succ q_2''$, $p_i' \vee q_1'' \vee q_2'' \sim p_i' \succ q_3''$ etc., and the equivalence in (16) is seen. Because of (16), Lemma 4-16 applies to $p_1' \leqq \cdots \leqq p_n'$, $P_1 \leqq \cdots \leqq P_n$, with the result that for a partial isometry v with $vv^* = 1$

$$vp_i'v^* \geqq P_i \quad \text{for all } i\,, \quad \text{and} \quad v^*v \geqq p_1'\,. \tag{17}$$

Then, $vq_j'v^*$ is a projection for every index j, and

$$vq_j'v^* \sim q_j \sim q_j \sim q_j'' . \tag{18}$$

From (16) and (17) we see that $vp_i'v^* \geqq P_1 = p_1' \vee q \geqq q$, and due to $vp_1'v^* \geqq vq_m'v^*$ we obtain: $vp_1'v^* \geqq q \vee vq_m'v^* = Q$. Look at the finite W^*-algebra $Q\mathcal{M}Q$. By assumption on ω, $\omega|_{Q\mathcal{M}Q}$ is a normal positive linear functional on $Q\mathcal{M}Q$, so Lemma 5-5 applies with the result, that there is $w \in \mathcal{M}$, with $w^*w = Q = ww^*$, and

$$\omega\Big(w\Big(\sum_j vq_j'v^*\Big)w^*\Big) \geqq \sum_j K_Q(\omega, vq_j'v) - \varepsilon , \tag{19}$$

where K_Q means the Ky Fan functional on $Q\mathcal{M}Q$. By definition of q and Q, we see q_j'', $vq_j'v^* \in Q\mathcal{M}Q$. Hence $vq_j'v^* \sim q_j''$ in $Q\mathcal{M}Q$. From principle Chapter 4, (13), it comes that $K_Q(\omega, vq_j'v^*) = K_Q(\omega, q_j'')$, and together with (15) we may conclude that

$$K_Q(\omega, vq_j'v^*) \geqq \omega(q_j'') \geqq K(\omega, q_j) - \varepsilon . \tag{20}$$

Taking (19) together with (20), we can derive that

$$\omega\Big(w\Big(\sum_j vq_j'v^*\Big)w^*\Big) \geqq \sum_j K(\omega, q_j) - (m+1)\,\varepsilon .$$

By defining $w' = w + Q^\perp$, we extend w to a unitary element in $\mathcal{M}$, and due to $vq_m'v^* \leqq Q$ we arrive at

$$\omega\Big(w'\Big(\sum_j vq_j'v^*\Big)w'^*\Big) \geqq \sum_j K(\omega, q_j) - (m+1)\,\varepsilon . \tag{21}$$

Because of $vp_1'v^* \geqq Q$ one remarks that $w'vp_i'v^*w'^* = vp_i'v^*$ for all i. The latter, together with (13), (14), (16) and (17), implies that

$$\sum_i \omega(w'vp_i'v^*w'^*) \geqq \sum_i K(\omega, p_i) - \varepsilon . \tag{22}$$

Define a partial isometry b as $b = w'vu$. Then, (13), (21) and (22) guarantee the validity of

$$\omega\Big(b\Big(\sum_i p_i + \sum_j q_j\Big)b^*\Big) \geqq \sum_j K(\omega, p_i) + \sum_j K(\omega, q_j) - (m+2)\,\varepsilon .$$

Due to Lemma 3-4 and $b \in \mathcal{E}$, we then see the validity of

$$K\Big(\omega, \sum_i p_i + \sum_j q_j\Big) \geqq \sum_j K(\omega, q_i) + \sum_j K(\omega, q_j) - (m+2)\,\varepsilon .$$

The preceding inequality, however, has to hold for $\varepsilon > 0$ as small as we like. This, together with subadditivity of the Ky Fan functional, requires that

$$K\Big(\omega, \sum_i p_i + \sum_j q_j\Big) = \sum_i K(\omega, p_i) + \sum_j K(\omega, q_j) ,$$

for the normal, positive linear form ω. By our remarks from the beginning of this proof, Lemma 5-6 is then justified. ▮

Now, let $\mathcal{M}$ be a general W^*-algebra. Then, $\mathcal{M}$ is the direct sum of three W^*-algebras which are finite, properly infinite and semi-finite, and purely infinite, respectively. Then, by Lemma 5-3, 5-5, 5-6 we are sure that Lemma 5-2 holds, on a general W^*-algebra. Due to the implication (2) we then have that the Σ-property is universal, so Theorem 5-1 is proved. ▮

5.5. A duality theorem, remarks

In extracting the Σ-property, we have arrived at a point which allows us to realize the deep connection between the two $\succ$-relations, the one defined in the projection lattice $\mathcal{M}^p$, and the other one acting in the state space $\mathcal{S}$ of the W^*-algebra under discussion. The result reads as (see [9, 13, 118]):

Theorem 5-7 (Duality theorem). *Let $\mathcal{M}$ be a W^*-algebra, with projection lattice $\mathcal{M}^p$ and state space $\mathcal{S}$. Let $p, q \in \mathcal{M}^p$, and $\omega, \nu \in \mathcal{S}$. Then,*

$$p \succ q \quad \text{if and only if} \quad K(\varrho, p) \geqq K(\varrho, q) \quad \text{for all} \quad \varrho \in \mathcal{S};$$

$$\omega \succ \nu \quad \text{if and only if} \quad K(\omega, Q) \leqq K(\nu, Q) \quad \text{for all} \quad Q \in \mathcal{M}^p .$$

The proof is a direct consequence of Definition 5-1, Theorem 5-1, Theorem 3-8 and Theorem 4-6. ∎

The formulated result shows that the relation $\succ$ on $\mathcal{S}$, the order structure is based on, is in case of a W^*-algebra in a well-defined sense the dual of v. Neumann's $\succ$ on the projection lattice.

We remark that Theorem 3-8, which is an immediate consequence of the Σ-property, can be generalized.*) Indeed, almost any C^*-algebra of physical relevance is quasi-local, with the local algebras being W^*-algebras. For such algebras, however, it is easy to see the implication $K(\omega, Q) \leqq K(\nu, Q)$ for all $Q \in \mathcal{A}$, projections, $\rightarrow \omega \succ \nu$ to be valid for states ω, ν on the C^*-algebra in question.

5.6. Positive linear forms on C^*-algebras and the Σ-property

In this section we shall investigate whether or not the Σ-property can be generalized in a suitable way (see [6, 118]).

Let $\mathcal{A}$ denote a C^*-algebra with unity $\mathbf{1}$. By $\mathcal{A}^{**}$ the second dual of $\mathcal{A}$ is meant. As usual, we may think of $\mathcal{A}$ as canonically embedded into $\mathcal{A}^{**}$, and $\mathcal{A}$ will be identified with its embedding. We remember that $\mathcal{A}^*$ then can be identified in a canonical manner with $(\mathcal{A}^{**})_*$, the predual of $\mathcal{A}^{**}$. Whenever $\omega \in \mathcal{A}^*_+$, the normal positive linear form on $\mathcal{A}^{**}$ uniquely extending ω, will be denoted by the same letter. Now, let K^{**} be the Ky Fan functional with respect to $\mathcal{A}^{**}$. Assume $a \in \mathcal{A}_+$, and $\omega \in \mathcal{A}^*_+$. Then, because of Lemma 3-4 we have in analogy with 5.1., (7):

$$K(\omega, a) = K^{**}(\omega, a) \quad \text{for all} \quad a \in \mathcal{A}_+ \quad \text{and for all} \quad \omega \in \mathcal{A}^*_+ . \tag{1}$$

In the following, relation (1) will be important, for it will enable us to insert the Σ-property with respect to $\mathcal{A}^{**}$ in order to see what it means on a C^*-algebra. The latter requires us to rewrite the Σ-property on a W^*-algebra into an equivalent form avoiding the usage of the item "projection".

To this end, let $\mathbf{F}$ denote the set of (finite) real valued continuous, monotonously increasing functions over the reals. Then, for $g \in \mathbf{F}$ and $a \in \mathcal{A}_h$ we let $g(a)$ be defined by the functional calculus in C^*-algebras.

*) Compare this to the remark concerning AW^*-algebras made in the introduction.

Lemma 5-8. *Let $\mathcal{M}$ be a W^*-algebra. Then the Σ-property is equivalent to*

$$K(\omega, g(a) + f(a)) = K(\omega, g(a)) + K(\omega, f(a)) \tag{2}$$

for every $\omega \in \mathcal{M}_+^*$, $a \in \mathcal{M}_+$ and $g, f \in \mathbf{F}$.

Proof. Let $\{e(t)\}$ denote the resolution of the identity (continuous from the left), corresponding to a. Then, $\omega_t = \omega(e(t)^\perp)$ and $\omega_t' = K(\omega, e(t)^\perp)$ are monotonously decreasing, and both

$$\int_0^\infty \omega_t\, dt\,, \qquad \int_0^\infty \omega_t'\, dt\,,$$

exist in the usual sense. It is plain to see (use approximations of a as in Chapter 4, (28), (29)) that

$$\omega(a) = \int_0^\infty \omega_t\, dt\,, \qquad K(\omega, a) = \int_0^\infty \omega_t'\, dt\,,$$

where the Σ-property in its usual formulation has been inserted. Now, one easily checks that, for $h \in \mathbf{F}$, the following holds:

$$\omega(h(a)) = \int_0^\infty \omega_t h(dt) + h(0)\,\omega(1)\,,$$

and — as a consequence from the latter — we see that

$$K(\omega, h(a)) = \int_0^\infty \omega_t' h(dt) + h(0)\,\omega(1)\,. \tag{3}$$

Let $f, g \in \mathbf{F}$. Then, from (3) and convexity of $\mathbf{F}$ we see that we are allowed to conclude as follows:

$$\begin{aligned} K(\omega, g(a) + f(a)) &= \int_0^\infty \omega_t'(g + f)\,(dt) + (g + f)\,(0)\,\omega(1) \\ &= \int_0^\infty \omega_t' g(dt) + g(0)\,\omega(1) + \int_0^\infty \omega_t' f(dt) + f(0)\,\omega(1) \\ &= K(\omega, g(a)) + K(\omega, f(a))\,, \end{aligned}$$

and one direction of the assertion is established.

To verify the reverse, assume (2) is valid for every $\omega \in \mathcal{M}_+^*$, for all $a \in \mathcal{M}_+$, and for each pair $g, f \in \mathbf{F}$. Let $p_1 < \cdots < p_n$ be projections of $\mathcal{M}$. We define projections q_k by $q_n = p_1$, and $q_k = p_{n-k+1} - p_{n-k}$ for $k = 1, \ldots, n-1$. Setting $a = \sum_{k=1}^{n-1} k q_k$, and defining $g, g_k \in \mathbf{F}$ by

$$g = \sum_k g_k\,, \qquad g_k(t) = \begin{cases} 0 & \text{for } t \leq k - 1; \\ 1 & \text{for } t \geq k; \\ t - k + 1 & \text{otherwise};\,, \end{cases}$$

one easily recognizes the relation $g_k(a) = \sum_{i=k}^{n} q_i = p_{n-k+1}$. Therefore, we may write $\sum_i p_i = \sum_{k=1}^{n} g_k(a) = g(a)$. The latter, together with an obvious consequence of (2) and definitions from above, gives

$$K(\omega, \textstyle\sum p_i) = K\left(\omega, \sum_k g_k(a)\right) = \sum_k K(\omega, g_k(a)) = \sum_k K(\omega, p_k)\,,$$

and the Σ-property appears as a consequence of (2). ∎

Let us continue in our considerations concerning the C^*-algebra $\mathcal{A}$. Let $g \in \mathbf{F}$. Then, by the spectral mapping theorem and (1), we see the validity of $K(\omega, g(a)) = K^{**}(\omega, g(a))$ for all $a \in \mathcal{A}_+$ and every $\omega \in \mathcal{A}_+^*$. The right-hand side of the preceding equality, however, refers exclusively to a W^*-algebra, so Lemma 5-8 applies and results in (see [6, 13]):

Theorem 5-9 (Σ-property in C^*-algebras). *Let $\mathcal{A}$ be a C^*-algebra with* **1**, *and* $\mathbf{F}$ *be the set of real valued continuous, monotonously increasing functions on the reals. Then, for every $\omega \in \mathcal{A}_+^*$ we have*

$$K(\omega, g(a) + f(a)) = K(\omega, g(a)) + K(\omega, f(a))$$

whenever $g, f \in \mathbf{F}$ and $a \in \mathcal{A}_+$.

Obviously, the a in our formulation can be thought of as being replaced with an arbitrary *Hermitian* element without violating equality.

Let us now give two (more or less trivial) examples for an application of Theorem 5-9.

Example 5-10. Let $\mathcal{A}$ be a C^*-algebra with **1**, H an Hermitian element of $\mathcal{A}$. Assume ω is a *passive* state with respect to the dynamic generated by H (see Pusz and Woronowicz [67]). Then, ω is also passive with respect to every dynamic generated by a Hamiltonian of the form $g(H)$ provided $g \in \mathbf{F}$.

Proof. Passivity means that $\omega(H) \leqq \omega^u(H)$ for every $u \in \mathcal{U}$ (this formulation is equivalent to the original one). In defining $a = \mathbf{1} - H$, we have $\omega(a) = K(\omega, a)$.

By (1) we then also have $\omega(a) = K^{**}(\omega, a)$. Let ω_t, ω_t' be defined as in the proof of Lemma 5-8, for $a \in \mathcal{A}^{**} = \mathcal{M}$. It means no restriction if we assume $a \geqq o$. Then, from the passivity condition we get

$$\int_0^\infty (\omega_t' - \omega_t)\, dt = K^{**}(\omega, a) - \omega(a) = 0\,.$$

Because of $\omega_t' \geqq \omega_t$ for all t, we have to conclude that $\omega_t' = \omega_t$ a.e. Let $g \in \mathbf{F}$. Then

$$\omega(g(a)) - K^{**}(\omega, g(a))$$
$$= \int_0^\infty (\omega_t' - \omega_t)\, g(dt) + (g(0)\,\omega(\mathbf{1}) - g(0)\,\omega(\mathbf{1})) = 0\,.$$

In applying (1) we arrive at $\omega(g(a)) = K(\omega, g(a))$. Thus ω is passive with respect to $\mathbf{1} - g(a)$ with arbitrarily chosen $g \in \mathbf{F}$. $\mathbf{F}$ containing the constant functions makes that: if $f(t) = t + 1$, then ω is passive with respect to $\mathbf{1} - (fg)(a) = 1 - g(a) - \mathbf{1} = -g(a)$. Defining $b(t) = 1 - t$, we have $a = b(H)$ and, because of $(-g)\,b \in \mathbf{F}$ for all $g \in \mathbf{F}$, we obtain:

$$\mathbf{1} - g(a) = ((-g)\,b)\,(H) = r(H)\,, \quad \text{with} \quad r \in \mathbf{F}\,.$$

Because of $b^2(t) = t$, however, r may stand for an arbitrary element m of $\mathbf{F}$ (one takes $g = -mb$, then $g \in \mathbf{F}$ and $r = (-g)\,b = mbb = mb^2 = m$). This concludes the proof. ∎

Example 5-11. Let $P(t) = \sum_{k=0}^{n} r_k t^k$ be a real polynomial, the derivative $P'(t)$ of which is non-negative for $t \in [0, 1]$. Then, for every a taken from $o \leqq a \leqq 1$:

$$K(\omega, P(a)) = \sum_{k=0}^{n} r_k K(\omega, a^k) \quad \text{for all} \quad \omega \in \mathcal{A}_+^* .$$

Proof. Because of $\sum_{r_k \geqq 0} r_k t^k = P(t) + \sum_{r_k < 0} |r_k| \, t^k$, and since t^k, $P(t)$ over $[0, 1]$ may be thought of as parts of functions from F, Theorem 5-9 applies and the result follows. ∎

6. The dual structure in W^*-algebras

6.1. Introduction

The aim of this chapter will be to introduce and investigate a further partial ordering, which is now defined in the Hermitian (resp. positive) portion $\mathcal{M}_h$ of a general W^*-algebra $\mathcal{M}$. We introduce $(\mathcal{M}_h, \succ)$ as:

Definition 6-1. Let $\mathcal{M}$ be a W^*-algebra. Then, $a \succ b$ for $a, b \in \mathcal{M}_h$ if $K(\omega, a) \geqq K(\omega, b)$ for all $\omega \in \mathcal{M}_+^*$.

In remembering Theorem 4-6 or Theorem 5-7 we become aware of the fact that $\succ$ reduces to v. Neumann's relation if both, a, b, belong to the projection set $\mathcal{M}^p$. Thus, the so introduced relation $\succ$ on $\mathcal{M}_h$ provides a further possibility for extending $\succ$ from $\mathcal{M}^p$ to more general sets of operators.

Since $\succ$ on $\mathcal{M}_h$ is fully determined when restricted to the positive cone $\mathcal{M}_+$, we rely upon the things we let follow when considering $(\mathcal{M}_+, \succ)$.

In case of $(\mathcal{S}, \succ)$ we insulated a certain "minimal" set of positive operators — the projections $\mathcal{M}^p$ — that totally characterize the relation via the Ky Fan functionals and show a very simple structure (cf. Theorem 5-7). Similarly, one goal of our investigation should be to find a certain "minimal" set of states (or positive linear functionals) which are sufficient to characterize $(\mathcal{M}_+, \succ)$ via the Ky Fan functionals. Like the projections in $\mathcal{M}_+$, the members of this distinguished minimal set of functionals should exhibit a "canonical" structure, and one should wonder whether or not a dual Σ-property could be formulated in this case, too. Finally, similarly to the situation we met for $(\mathcal{S}, \succ)$, we aim at a formulation of the partial order in question which does not make reference to the notion of state, i.e. we have to look for an intrinsic characterization.

At last, keeping in mind that the relations between $\mathcal{M}$ and $\mathcal{M}^*$ are not totally symmetric ones, in case of $(\mathcal{M}_+, \succ)$ we cannot expect to find such a comprehensive relation like $\omega \in w^*\text{-cl conv }\{\varrho^u\}_{u \in \mathcal{U}}$ for $\omega \succ \varrho$ as it occurs in case of $(\mathcal{S}, \succ)$. In advance, let us formulate one result of our considerations in form of (see [118]):

Theorem 6-1. *Let $\mathcal{M}$ be a W^*-algebra, with centre $\mathcal{Z}$. There exists a family $\{\mathcal{J}_q\}_{q \in \mathcal{M}^p}$ of maps from $\mathcal{M}_+$ into $\mathcal{Z}_+$, with the following properties:*

1. $\|\mathcal{J}_q(a)\| \leqq \|a\|$, $\quad a \in \mathcal{M}_+$;
2. $\mathcal{J}_q(1) = \mathcal{J}_p(q) = c(q)$ $\quad$ *(the central support of q)*;
3. $\mathcal{J}_q(b^*b) = \mathcal{J}_q(bb^*)$, $\quad b \in \mathcal{M}$;
4. $\mathcal{J}_q = \mathcal{J}_p$ *for* $p \sim q$;
5. $\mathcal{J}_q$ *is subadditive on* $\mathcal{M}_+$;

6. $\mathcal{J}_q(z.) = z\mathcal{J}_q(.)$, $\quad z \in \mathcal{Z}_+$;
7. $\mathcal{J}_q$ *is monotonously increasing on* $\mathcal{M}_+$;
8. $\mathcal{J}_q\left(\sum_i q_i\right) = \sum_i \mathcal{J}_q(q_i)$ *for every finite ordered sequence* (q_i) *of projections* (Σ*-property of* $\mathcal{J}_q$);
9. $a \succ b$ *for* $a, b \in \mathcal{M}_+$ *iff* $\mathcal{J}_q(a) \geqq \mathcal{J}_q(b)$ *for every* $q \in \mathcal{M}^p$.

From the listed points in Theorem 6-1 it is easily read off that the maps of the form $\mathcal{J}_q(.)\, q$ provide examples of centre-valued convex traces with respect to the W^*-algebra $q\mathcal{M}q$, for every projection q taken from $\mathcal{M}$. And this is not at all accidental.

6.2. Technical preliminaries

In this part some useful technical facts will be concentrated; to some extent one might think of the presented matter as a continuation of 4.5.

Let $\mathcal{M}$ be a W^*-algebra, and define $\mathcal{M}_+^0$ as the set of those elements of $\mathcal{M}_+$, which have spectra consisting of finitely many points exclusively. Then, for $a \in \mathcal{M}_+$ let us define sets $L(a)$ and $U(a)$ by $L(a) = \{b \in \mathcal{M}_+^0 : b \leqq a\}$, $U(a) = \{b \in \mathcal{M}_+^0 : b \geqq a\}$. We then have:

Remark 6-2. Assume $a, b \in \mathcal{M}_+$. $a \succ b$ iff $a' \succ b'$ for every $a' \in U(a)$, $b' \in L(b)$.

Proof. For every $d \in \mathcal{M}_+$, one has $d \in \|\text{-cl}\, L(d) \cap \|\text{-cl}\, U(d)$. The latter, together with uniform continuity and monotonicity of Ky Fan functionals on $\mathcal{M}_+$, gives the validity of the assertion. ∎

Let $\{z_i\}$ be a finite orthogonal decomposition of $\mathbf{1}$ into central projections z_i. Then, in consequence of Definition 6-1, we see $a \succ b$ for $a, b \in \mathcal{M}_+$ iff $az_i \succ bz_i$ for every i. Let $a, b \in \mathcal{M}_+^0$, i.e. $a = \sum_j \alpha_j p_j$ and $b = \sum_i \beta_i q_i$, with finitely many non-negative reals α_j, β_i and sequences $q_1 < \cdots < q_n = \mathbf{1}$ and $p_1 < \cdots < p_m = \mathbf{1}$ of projections.

By the same sort of arguments we stressed at the beginning of the proof of Lemma 5-4 we provide ourselves with a finite orthogonal decomposition of $\mathbf{1}$ into central projections, $\{z'_k\}$, such that $\{p_s z'_k, q_r z'_k\}_{s,r}$ (s, r are the running indices) consists of mutually comparable projections in the sense of $\succ$. We consider the finite systems $\{p_s z'_k\}_s$ and $\{q_r z'_k\}_r$ of projections. We may be sure to find finite orthogonal decompositions $\{z_{1kj}\}_j$ and $\{z_{2kl}\}_l$ of $\mathbf{1}$ into central projections such that $\{p_s z'_k z_{1kj}\}_s$ and $\{q_r z'_k z_{2kl}\}_r$ both contain only properly infinite or finite projections (one has to put arguments as in the proof of Lemma 5-6). Finally, Lemma 4-15 guarantees the existence of finite orthogonal decompositions $\{z_{1kjf}\}_f$ and $\{z_{2klt}\}_t$ of $\mathbf{1}$ into central projections such that for $\{p_s z'_k z_{1kj} z_{1kjf}\}_s$ and $\{q_r z'_k z_{2kl} z_{2klt}\}_r$ the following holds:

(a) only finite or properly infinite projections belong to the systems in question,

(b) if $p_s z'_k z_{1kj} z_{1kjf}$ is properly infinite, so either $p_s z'_k z_{1kj} z_{1kjf} \sim p_{s+1} z'_k z_{1kj} z_{1kjf}$, or $p_s z'_k z_{1kj} z_{1kjf} \in \mathcal{J}_{p_{s+1} z'_k z_{1kj} z_{1kjf}}$ (c-ideal), and an analogous condition holds for properly infinite members of $\{q_r z'_k z_{2kl} z_{2klt}\}_r$.

Now, let us form the coarsest finite orthogonal decomposition of $\mathbf{1}$ into central projections which is finer than all the central decompositions considered above. Call this decomposition $\{z_i\}$. Then, one easily checks that each of the systems $\{p_s z_i, q_r z_i\}_{s,r}$ inherits all the nice properties mentioned above. In remembering the beginning of our discussion, we then have it that $a \succsim b$ is fully determined if the meaning of $\succsim$ between elements az_i, bz_i can be clarified, where the $\{p_s z_i\}_s$, $\{q_r z_i\}_r$ are in the canonical relationship to each other that we have considered. In this sense we have:

Remark 6-3. The structure of $\succsim$ on $\mathcal{M}_+$ is fully reduced to the question whether $a \succsim b$ is valid in case of $a, b \in \mathcal{M}_+$ with the following properties: $a = \sum_j \alpha_j p_j$, $b = \sum_i \beta_i q_i$, with $\alpha_j, \beta_i \geqq 0$ and $o < p_1 < \cdots < p_n = z$, $o < q_1 < \cdots < q_m = z$ projections, such that $z \in \mathcal{Z}^p$ and:

$\{p_j\}$ and $\{q_i\}$ contain only finite or properly infinite projections; (1)

every p_j is comparable to each q_i in the sense of $\succsim$; (2)

if p_j (resp. q_i) is properly infinite, so either $p_j \sim p_{j+1}$ (resp. $q_i \sim q_{i+1}$) or $p_j \in \mathcal{J}_{p_{j+1}}$ (resp. $q_i \in \mathcal{J}_{p_{i+1}}$), where $\mathcal{J}_{p_{j+1}}$ means the c-ideal corresponding to $p_{j+1}\mathcal{M}p_{j+1}$. (3)

Lemma 6-4. *Let* $p_{01} < \cdots < p_{0n_0} < p_{11} < \cdots < p_{1n_1} < p_{21} < \cdots < p_{m-1,n_{m-1}} < p_{m1} < \cdots < p_{mn_m} = \mathbf{1}$ *be a sequence of projections with:*

(a) p_{0i} *is finite for all* i;

(b) p_{ir} *is properly infinite for* $i > 0$;

(c) $p_{ir} \sim p_{il}$ *for* $i \neq 0$ *and* $r, l \leqq n_i$;

(d) $p_{in_i} \in \mathcal{J}_{p_{i+1}, n_{i+1}}$ *for all* i.

Then there is a partial isometry $V \in \mathcal{M}$ *with* $VV^* = \mathbf{1}$ *such that*

(a′) $Vp_{0i}V^* = p_{0i}$ *for all* i,

(b′) $Vp_{ik}V^* = p_{in_i}$ *for all* $i > 0$, *and* $1 \leqq k \leqq n_i$.

Proof. By Lemma 4-13, (d) implies that $p_{kn_k} - p_{k-1,n_{k-1}} \sim p_{kn_k}$. Respecting (b) and (c), the same argument leads to $p_{k1} - p_{k-1,n_{k-1}} \sim p_{k1} \sim p_{kn_k}$. Hence, we have a partial isometry $v_k \in \mathcal{M}$ such that $v_k^* v_k = p_{k1} - p_{k-1,n_{k-1}} = q_k$ and $v_k v_k^* = p_{kn_k} - p_{k-1,n_{k-1}} = q'_k$, for $k = 1, 2, \ldots, m$. Define $v_0 = p_{0n_0} = q_0 = q'_0$. Because of $q_k q_l = o = q'_k q'_l$ in case of $k \neq l$, and $\sum_k q'_k = \mathbf{1}$, we recognize $V = \sum_k v_k$ to be a partial isometry obeying $VV^* = \mathbf{1}$. Finally, from the construction of V it follows the validity of (a′) and (b′). ∎

Lemma 6-5. *Let* $p_{01} < \cdots < p_{0r} < p_1 < \cdots < p_n = \mathbf{1}$ *and* $q_{01} < \cdots < q_{0s} < q_1 < \cdots < q_m = \mathbf{1}$ *be sequences of projections with:*

(a) p_{0i}, q_{0i} *finite;*

(b) p_i, q_i *properly infinite;*

(c) $p_i \in \mathcal{J}_{p_{i+1}}$, $q_j \in \mathcal{J}_{q_{j+1}}$ *for all* i, j;

(d) *every projection of* $\{p_{0i}, p_j\}_{i,j}$ *is comparable to every projection of* $\{q_{0k}, q_l\}_{k,l}$ *in the sense of* $\succ$.

Then, there is a unitary $u \in \mathcal{U}$ *such that* $\{up_{0i}u^*, up_ju^*, q_{0k}, q_l\}_{i,j,k,l}$ *is a* $\leq$*-ordered system of projections, and* $p \nsim q$ *for members* p, q *of the system whenever* $p \neq q$.

Proof. Because of (d), there is a minimal set of projections (i.e. if p, q belong to the set in question and $p \neq q$, so $p \nsim q$) $Q_{01} < \cdots < Q_{0t} < Q_1 < \cdots < Q_f = 1$ such that

(a′) if $p \in \{p_{0i}, q_{0j}\}_{i,j}$, we find i, $1 \leq i \leq t$, with $Q_{0i} \sim p$;

(b′) if $p \in \{p_j, q_l\}_{j,l}$, there is i, $1 \leq i \leq f$, with $Q_i \sim p$.

We choose indices $i_1 < \cdots < i_n$ and $j_1 < \cdots < j_r$ in such a way that $Q_{i_k} \sim p_k$ and $Q_{0j_l} \sim p_{0l}$. Then, due to (c) and Lemma 4-14, we find a unitary w such that $wp_kw^* = Q_{i_k}$. Automatically, $wp_{0r}w^* \leq Q_{i_1}$ for all r. Since Q_{0j_r}, $wp_{0r}w^* \in \mathcal{J}_{Q_{i_1}}$ is true in virtue of Lemma 4-13, we see $Q_{i_1} - Q_{0j_r} \sim Q_{i_1}$ and $Q_{i_1} - wp_{0r}w^* \sim Q_{i_1}$. The latter implies $Q_{i_1} - Q_{0j_r} \sim Q_{i_1} - wp_{0r}w^*$. Since w is unitary, and $Q_{0j_l} \sim p_{0l}$, and these projections are finite, we can be sure (cf. SAKAI [80], 2.4.2.) that $Q_{0j_i} - Q_{0j_{i-1}} \sim wp_{0i}w^* - wp_{0,i-1}w^*$ is valid for all i. Therefore, there has to exist a unitary v in $Q_{i_1}\mathcal{M}Q_{i_1}$ transforming $wp_{0i}w^*$ into Q_{0j_i} simultaneously for every i. We define $v' = v + Q_{i_1}^{\perp}$. Then, $v' \in \mathcal{U}$ and $u_1 = v'w \in \mathcal{U}$, with the property that $\{p_{0i}, p_j\}_{i,j}$ is mapped into a subset of $\{Q_{0i}, Q_j\}_{i,j}$ by applying $u_1 . u_1^*$. Analogously, we succeed in constructing a unitary u_2 such that $u_2 . u_2^*$ transforms $\{q_{0k}, q_l\}_{k,l}$ into a subset of $\{Q_{0i}, Q_j\}_{i,j}$. But then, $u = u_2^* u_1$ can be taken as the unitary in question. ∎

Lemma 6-6 (cf. [7]). *Let* $p \in \mathcal{M}^p$, *and assume* $\omega \in \mathcal{M}_+^*$. *Suppose* $q, q_0 \in \mathcal{M}^p$, *with* $q \sim q_0$, $q_0 \leq p$. *Then,* $K(\omega(p.p), q) = K_p(\omega, q_0)$, *where* K_p *denotes the Ky Fan functional with respect to* $p\mathcal{M}p$.

Proof. By our results of $\succ$ in the projection lattice of a W^*-algebra, and by Lemma 3-4 we may conclude as follows:

$$K(\omega(p.p), q) = K(\omega(p.p), q_0) = \sup_{a \in \mathcal{E}} \omega(papq_0pa^*p)$$
$$= \sup_{b \in \mathcal{E}_p} \omega(bq_0b^*) = K_p(\omega, q_0)\,. \quad ∎$$

6.3. The purely infinite case

Throughout this section $\mathcal{M}$ is assumed to be a W^*-algebra of type III. Among W^*-algebras, the purely infinite case is distinguished, since it allows a very comprehensive characterization of $(\mathcal{M}_+, \succ)$, see [118]):

Theorem 6-7. *Let* $\mathcal{M}$ *be a* W^**-algebra of type* III. *Then,* $b \succ a$ *for* $a, b \in \mathcal{M}_+$ *if and only if there exists a sequence* $\{w_n\}$ *of partial isometries such that* $w_nw_n^* = 1$ *and* $n^{-1}1 \geq a - w_nbw_n^*$ *for all* n.

Proof. The conditions are sufficient for $b \succ a$. Indeed, because of $\|w_n\| \leqq 1$ we might conclude as follows:

$$\begin{aligned}\forall \omega \in \mathcal{M}_+^*, \quad K(\omega, a) &\leqq K(\omega, n^{-1}\mathbf{1} + w_n b w_n^*) = n^{-1}\omega(\mathbf{1}) + K(\omega, w_n b w_n^*) \\ &\leqq n^{-1}\omega(\mathbf{1}) + \sup_{d \in \mathcal{E}} \omega(d w_n b w_n^* d^*) \\ &\leqq n^{-1}\omega(\mathbf{1}) + \sup_{c \in \mathcal{E}} \omega(c b c^*) \\ &= n^{-1}\omega(\mathbf{1}) + K(\omega, b)\end{aligned}$$

for every n, so $K(\omega, a) \leqq K(\omega, b)$, $\omega \in \mathcal{M}_+^*$, i.e. $b \succ a$.

The formulated conditions are necessary for $b \succ a$.

To prove this we have to make reference to the main results of the Chapters 4 and 5. Let $b \succ a$. Then, we can choose $b_n \in U(b)$ and $a_n \in L(a)$ such that $(2n)^{-1}\mathbf{1} \geqq b_n - b \geqq o$, $(2n)^{-1}\mathbf{1} \geqq a - a_n \geqq o$ (cf. Remark 6-2). Fix n. Then, by arguments we used in deriving Remark 6-3, we may be assured of the existence of a finite orthogonal decomposition $\{z_i\}$ of $\mathbf{1}$ into central projections such that $X_i = a_n z_i$ and $Y_i = b_n z_i$ satisfy conditions (1)—(3) (see Remark 6-3). We fix i, and write $X_i = \sum_j x_j q_j$ and $Y_i = \sum_l y_l p_l$, with x_j, $y_l \geqq 0$ and sequences of projections $o < p_1 < \cdots < p_n = z_i$ and $o < q_1 < \cdots < q_m = z_i$. By Lemma 6-4, we can provide us with partial isometries V, $W \in \mathcal{M}$ such that $VV^* = z_i = WW^*$ and

$$X_i \leqq VX_iV^* = X_i' = \sum_j x_j' q_j' , \qquad Y_i \leqq WY_iW^* = Y_i' = \sum_l y_l' p_l' . \tag{4}$$

Here x_j', $y_l' \geqq 0$ are reals, $o < q_1' < \cdots < q_r' = z_i$ and $o < p_1' < \cdots < p_s' = z_i$ are subsequences of $\{q_j\}$ and $\{p_l\}$ respectively, with $q_j' \in \mathcal{J}_{q'_{j+1}}$ and $p_l' \in \mathcal{J}_{p'_{l+1}}$ (respectively) for all the possible j, l. The latter and Lemma 6-5 guarantee the existence of unitary $u \in \mathcal{M}z_i$ such that $\{Q_k\}_{k=1}^N = \{uq_j'u, p_l'\}_{j,l}$ is ordered: $o < Q_1 < \cdots < Q_N = z_i$. Then, we find reals (0 is included) t_k, $t_k' \geqq 0$ such that

$$X_i'' = uX_i'u^* = \sum_k t_k Q_k \quad \text{and} \quad Y_i' = \sum_k t_k' Q_k . \tag{5}$$

From (4), (5) and Lemma 3-4 one has to follow

$$K(\omega, X_i'') = K(\omega, X_i') = K(\omega, X_i) , \qquad \text{for all } \omega \in \mathcal{M}_+^* , \tag{6}$$

and

$$K(\omega, Y_i') = K(\omega, Y_i) , \qquad \text{for all } \omega \in \mathcal{M}_+^* . \tag{7}$$

By assumptions on a, b, a_n, b_n, and taking into account the definitions of X_i, Y_i, we see for any $\omega \in \mathcal{M}_+^*$:

$$K(\omega, X_i) \leqq K(\omega, Y_i) . \tag{8}$$

From (8), however, together with (6) and (7) we conclude the validity of

$$K(\omega, X_i'') \leqq K(\omega, Y_i') , \qquad \text{for all } \omega \in \mathcal{M}_+^* . \tag{9}$$

In using representation (5) and the Σ-property, we get from (9)

$$\sum_k t_k K(\omega, Q_k) \leqq \sum_k t_k' K(\omega, Q_k) \qquad \text{for every } \omega \in \mathcal{M}_+^* . \tag{10}$$

Especially, (10) has to be valid for $\omega_k \in \mathcal{M}_+^*$ defined by:

$$\omega_k = \varrho_k(z_k' Q_k \cdot Q_k z_k') , \qquad \text{with} \quad \varrho_k \in (Q_k \mathcal{M} Q_k z_k')_+^* \tag{11}$$
$$\text{such that } \varrho_k(\mathcal{J}_{Q_k z_k'}) = (0) \quad \text{and} \quad \varrho_k(Q_k z_k') = 1 ,$$

where the central projection z_k' has been chosen in such a manner that $Q_{k-1}z_k' \in \mathcal{J}_{Q_k z_k'}$ (this is always possible since u was chosen in accordance with Lemma 6-5) for every index k.

From Lemma 6-6 and (11) then follows that

$$K(\omega_k, Q_j) = \begin{cases} 1 & \text{for } j \geqq k, \\ K_{Q_k z_k'}(\varrho_k, Q_j z_k') = 0 & \text{for } j < k, \end{cases} \tag{12}$$

because $Q_j z_k' \in \mathcal{J}_{Q_k z_k'}$ for every $j < k$. But then, from (10) and (12) we see that

$$\sum_{j \geqq k} t_j \leqq \sum_{j \geqq k} t_j' \quad \text{for} \quad k = 1, 2, \ldots, N. \tag{13}$$

In virtue of representation (5) we may rewrite X_i'' and Y_i', respectively, into the form

$$X_i'' = \sum_{k=1}^{N} \Big(\sum_{j \geqq k} t_j \Big) (Q_k - Q_{k-1}), \qquad Y_i' = \sum_{k=1}^{N} \Big(\sum_{j \geqq k} t_j' \Big) (Q_k - Q_{k-1}), \tag{14}$$

where $Q_0 = o$ by definition.

From (14), it comes that the system (13) of inequalities results in

$$X_i'' \leqq Y_i'. \tag{15}$$

By (4) and (5) we see that (15) means $uX_iu^* \leqq uX_i'u^* = X_i'' \leqq WY_iW^*$, so

$$X_i \leqq v_i Y_i v_i^*, \quad \text{with} \quad v_i = u^* W \quad \text{and} \quad v_i v_i^* = z_i. \tag{16}$$

We now lift the fixation on i, and arrive at an v_i for every index i. Because of (16), $w_n = \sum_i v_i$ is a partial isometry fulfilling $w_n w_n^* = 1 = \sum_i z_i$, and

$$a_n = \sum_i X_i \leqq \sum_i v_i Y_i v_i^* = w_n b_n w_n^*,$$

so $w_n(n^{-1}2 1 + b)\, w_n^* \geqq w_n b_n w_n^* \geqq a_n \geqq a - (2n)^{-1}1$. The latter, however, implies $n^{-1}1 \geqq a - w_n b w_n^*$ due to $w_n w_n^* = 1$.

Finally, nothing can hinder us from lifting the fixation on n, and the proof is complete. ▮

Consequence. In case of $b \succsim a$, $b, a \in \mathcal{M}_+$, the sequence of partial isometries $\{w_n\}$ in the assertion of Theorem 6-7 can be chosen in such a manner that

$$\text{uniform-}\lim_n \, (a w_n b w_n^* - w_n b w_n^* a) = o.$$

Proof. The assertion is implied by the construction above of the w_n. A further consequence is formulated in form of:

Lemma 6-8. *Let $\mathcal{M}$ be a W^*-algebra of type* III, *and $a, b \in \mathcal{M}_+$. Then, $b \succsim a$ if and only if there is a sequence $\{d_n\} \subset \mathcal{M}$ such that $b \geqq d_n d_n^*$ and $d_n^* d_n \geqq a - n^{-1}1$, for every n.*

Proof. Sufficiency follows from $K(\omega, d_n^* d_n) = K(\omega, d_n d_n^*)$ for all n and for all $\omega \in \mathcal{M}_+^*$, and the monotonicity of $K(\omega, .)$. Necessity follows from Theorem 6-7 if we define $d_n = b^{1/2} w_n^*$ for all n, with w_n as in the assertion of Theorem 6-7. ▮

6.4. Approximation results

As it has been shown in the previous section, in case of a W^*-algebra of type III positive linear functionals of the following size play a decisive role in context of $(\mathcal{M}_+, \succsim)$ (cf. (11) in 6.3.):

$$\omega = \varrho(Q.Q), \quad \text{with} \quad Q \in \mathcal{M}^p \quad \text{and} \quad \varrho \in (Q\mathcal{M}Q)_+^*, \quad \varrho(\mathcal{J}_Q) = (0), \tag{17}$$

where $\mathcal{J}_Q$ means the c-ideal with respect to $Q\mathcal{M}Q$.

We will try to take (17) as an appropriate model for generalizations to arbitrary W^*-algebras.

Let $\mathcal{M}$ be a general W^*-algebra, and $Q \in \mathcal{M}^p$, $Q \neq o$. Then, by $\mathcal{S}_\infty(Q\mathcal{M}Q)$ we denote the set of maximally mixed states on the W^*-algebra $Q\mathcal{M}Q$ (cf. 4.4.), and we are going to extend states of $\mathcal{S}_\infty(Q\mathcal{M}Q)$ to states of the original algebra $\mathcal{M}$ in a straightforward manner by:

$$\mathcal{S}_\infty^Q = \{\omega \in \mathcal{S} : \exists \varrho \in \mathcal{S}_\infty(Q\mathcal{M}Q) \quad \text{with} \quad \omega = \varrho(Q.Q)\}. \tag{18}$$

Let $o < Q_1 < \cdots < Q_n = \mathbf{1}$ be a sequence of projections of $\mathcal{M}$. Then, a positive linear form ω on $\mathcal{M}$ will be said to be of canonical type with respect to $\{Q_i\}$ (see [13, 118]) ($\{Q_i\}$-canonical, for short) if there are non-negative reals $\{r_i\}$ and $\{\omega_i\} \subset \mathcal{S}_\infty^{Q_i}$ such that

$$\omega = \sum_i r_i \omega_i. \tag{19}$$

Throughout this section it will be our aim to establish some important properties of functionals of this form.

Two obvious properties are formulated as:

$$\text{the set of } \{Q_i\}\text{-canonical functionals is convex;} \tag{20'}$$

$$\text{if } \omega \text{ is } \{Q_i\}\text{-canonical, and } \omega' \text{ is } \{Q_j'\}\text{-canonical, so } \omega + \omega' \text{ is } \{Q_i, Q_j'\}\text{-canonical provided the system } \{Q_i, Q_j'\} \text{ is } \leqq\text{-ordered.} \tag{20''}$$

A first non-trivial property of canonical functionals reads as:

Lemma 6-9. *Let ω be $\{Q_j\}$-canonical. Then for every index j we have $K(\omega, Q_j) = \sum_i r_i K(\omega_i, Q_j) = \omega(wQ_jw^*)$, with a partial isometry w obeying $ww^* = \mathbf{1}$ and depending only of the system $\{Q_j\}$. Moreover, in the case of finite $\mathcal{M}$, $w = \mathbf{1}$ can be chosen (notations as in* (17) *and* (19)*).*

Proof. As it is now a standard procedure (cf. the reasoning of Remark 6-3), we may be assured of the existence of a finite orthogonal decomposition of $\mathbf{1}$ into central projections $\{z_j\}$ such that, for each j, the system $\{Q_i z_j\}_i$ of projections is either of the type needed for an application of Lemma 6-4 (with respect to the W^*-algebra $\mathcal{M}z_j$), or z_j is finite; in order to be non-trivial we assume at least one z_j to be properly infinite. In the case of j with z_j properly infinite, Lemma 6-4 applies, i.e. we find partial isometries v_j, $v_jv_j^* = z_j$, such that in $\{v_jQ_iv_j^*\}_i$ from $v_jQ_iv_j^* \neq v_jQ_{i+1}v_j^*$ always the relation $v_jQ_iv_j^* \nsim v_jQ_{i+1}v_j^*$ follows and

$$v_jQ_iv_j^* \geqq Q_iz_j, \quad \text{and equality in case of finite } Q_iz_j. \tag{21}$$

In case of z_j finite, however, we define $v_j = z_j$.

Let a partial isometry w be defined as $w = \sum_j v_j$. Fix j (for non-triviality, we shall assume j with properly infinite z_j).

Suppose $Q_i z_j$ is finite for $i \leqq k$, and properly infinite for $i > k$. Then, for $l \leqq k$

$$K(\omega_l, Q_i z_j) = \omega_l(w Q_i w^* z_j) = \omega_l(Q_i z_j) \quad \text{for all } i\,. \tag{22}$$

In fact, for $i \leqq l$

$$\begin{aligned} K(\omega_l, Q_i z_j) &= K\big(\varrho_l(Q_l z_j \cdot Q_l z_j), Q_i\big) = K_{Q_l z_j}(\varrho_l, Q_i z_j) \\ &= \varrho_l\big((Q_i z_j)^\S\big) = \varrho_l(Q_i z_j) = \omega_l(Q_i z_j) \\ &= \omega_l(w Q_i w^* z_j)\,, \end{aligned}$$

where $\omega_l = \varrho_l(Q_l \cdot Q_l)$, with $\varrho_l \in \mathcal{S}_\infty(Q_l \mathcal{M} Q_l)$, and Lemma 6-6, Theorem 4-12 (with § denoting the §-operation with respect to $Q_l \mathcal{M} Q_l$) have been inserted.

In the case of $i \geqq l$ (for $l \leqq k$) we have $Q_i \geqq Q_l$, so from

$$K(\omega_l, Q_i z_j) = K\big(\varrho_l(Q_l \cdot Q_l), Q_i z_j\big) \leqq \varrho_l(Q_l z_j) = \omega_l(z_j)$$

because of (21) we may draw the conclusion that

$$K(\omega_l, Q_i z_j) = \omega_l(Q_i Q_l z_j) = \varrho_l(Q_l z_j) = \omega_l(w Q_i w^* z_j)\,,$$

and (22) is seen to be true.

Let $l > k$. Then,

$$\begin{aligned} \omega_l(w Q_i w^* z_j) &= \omega_l(v_j Q_i v_j^*) = \varrho_l(Q_l v_j Q_i v_j^* Q_l z_j) \\ &= \varrho_l(Q_l z_j) = \omega_l(z_j) \quad \text{for} \quad Q_i z_j \succsim Q_l z_j \end{aligned} \tag{23}$$

(this especially holds for $i \geqq l$).

The latter follows from the choice of v_j. Now, since ϱ_l is in $\mathcal{S}_\infty(Q_l \mathcal{M} Q_l)$, and for $i < l$ with $Q_i z_j \not\succsim Q_l z_j$ by the choice of v_j follows that $w Q_i z_j w^* = v_j Q_i v_j^*$ belongs to the c-ideal $\mathcal{J}_{Q_l z_j}$, by means of Theorem 4-12 we see

$$\omega_l(w Q_i w^* z_j) = 0 \quad \text{for } i < l \quad \text{with } Q_i z_j \not\succsim Q_l z_j\,. \tag{24}$$

By (23) and (24), the whole i-range is covered. Assume i is such that (24) occurs. Then, stressing the arguments which led to (24) once more, in virtue of Lemma 6-6 we will arrive at

$$K(\omega_l, Q_i z_j) = K\big(\varrho_l(Q_l \cdot Q_l), Q_i z_j\big) = K_{Q_l z_j}(\varrho_l, Q_i z_j) = 0\,. \tag{25}$$

For those i's which correspond to (23), by Lemma 3-4 we may conclude

$$\omega_l(z_j) \geqq K(\omega_l, Q_i z_j) = \sup_{b \in \mathcal{E}} \omega_l(b Q_i b^* z_j) \geqq \omega_l(z_j)\,.$$

From the latter together with (23), (24) and (25), the equality

$$K(\omega_l, Q_i z_j) = \omega_l(w Q_i w^* z_j) \quad \text{for all } l > k \tag{26}$$

is obtained. Taking (23) together with (26) yields the validity of

$$K(\omega_l, Q_i z_j) = \omega_l(w Q_i w^* z_j) \quad \text{for all } l, i\,.$$

The latter equality resulting for every j, we may conclude to $K(\omega_l, Q_i) = \omega_l(w Q_i w^*)$ for all i, l. This, however, due to the structure of ω, leads to

$$\begin{aligned} K(\omega, Q_i) &\leqq \sum_l r_l K(\omega_l, Q_i) = \sum_l r_l \omega_l(w Q_i w^*) \\ &= \omega(w Q_i w^*) \leqq \sup_{b \in \mathcal{E}} \omega(b Q_i b^*) = K(\omega, Q_i)\,. \end{aligned}$$

Hence equality has to occur. Finally, in the case of finite $\mathcal{M}$ we have $w = 1$, and in the general case w depends only on $\{Q_i\}$ (especially, there is no correlation to the special ϱ_i's and r_i's occurring in the definition of ω). ∎

The just-derived property proves to be an essential ingredient for a proof of the following main result of this section (cf. [118]):

Proposition 6-10. *Let $\mathcal{M}$ be a W^*-algebra, and $o < Q_1 < \cdots < Q_n = \mathbf{1}$ be a sequence of projections of $\mathcal{M}$. Let $\varrho \in \mathcal{M}_+^*$. There is a $\{Q_i\}$-canonical functional $\omega \in \mathcal{M}_+^*$ such that $\omega(Q_i) = K(\omega, Q_i) = K(\varrho, Q_i)$ for each $i = 1, 2, \ldots, n$.*

To prepare for a proof of Proposition 6-10, we have to derive some results on finite W^*-algebras.

Lemma 6-11. *Let $\mathcal{N}$ be a finite W^*-algebra, $q_1 < \cdots < q_m = \mathbf{1}$ and $o = p_0 < p_1 < \cdots < p_n = \mathbf{1}$ projections of $\mathcal{N}$. Let $a = \sum_j \beta_j q_j$, with reals $\beta_j \geqq 0$. Assume $\omega \in \mathcal{N}_+^*$. Then, there exists a finite orthogonal decomposition of $\mathbf{1}$ into central projections, $\{z_i\}_{i=1}^k$, and a set $\{a_i\}_{i=1}^k \subset \mathcal{N}_+$ such that:*

(i) $\quad a_i = \sum_s \beta'_{is} p_s\,,\quad$ *with certain reals $\beta'_{is} \geqq 0$;*

(ii) $\quad (a_i p_l)^\natural z_i \geqq (a_i v p_l v^*)^\natural z_i \quad$ *for all l, i and for all $v \in \mathcal{U}$;*

(iii) $\quad \sum_i \omega\big(z_i (a_i p_l)^\natural\big) = K\big(\omega((a.)^\natural), p_l\big) \quad$ *for all l.*

Proof. Let $\{z_i\}_{i=1}^k$ and $u \in \mathcal{U}$ be chosen in accordance with Lemma 5-4. Then, for every i, $\{q_j z_i, u p_l u^* z_i\}_{j,l}$ is $\leqq$-ordered. We define:

$$q_j z_i = q_{ji}\,, \qquad u p_l u^* z_i = p_{li}\,. \tag{27}$$

Fix i. Since $\{q_{ji}, p_{li}\}_{j,l}$ is $\leqq$-ordered, there is s_j such that

$$p_{s_j i} \leqq q_{ji} \leqq p_{s_j+1,\,i}\,. \tag{28}$$

From the monotonicity of the operation $\natural$ (the centre-valued trace) together with (28) comes that t_{ji}, with $0 \leqq t_{ji} \leqq 1$, exists such that

$$t_{ji}\omega(p^\natural_{s_j i}) + (1 - t_{ji})\,\omega(p^\natural_{s_j+1,\,i}) = \omega(q^\natural_{ji})\,. \tag{29}$$

We let a_{ji} be defined as

$$a_{ji} = t_{ji} p_{s_j} + (1 - t_{ji})\, p_{s_j+1}\,. \tag{30}$$

Then, from the construction of a_{ji}, and because of the choice of u in accordance with Lemma 5-4, we get

$$(a_{ji} u^* p_{li} u)^\natural \geqq (a_{ji} v p_{li} v^*)^\natural \quad \text{for all } v \in \mathcal{U}\,. \tag{31}$$

Taking together (29), (30) and (31) gives to us

$$\omega\big((a_{ji} u^* p_{li} u)^\natural\big) = \begin{cases} \omega(p^\natural_{li}) & \text{for } l \leqq s_j\,, \\ \omega(q^\natural_{ji}) & \text{for } l > s_j\,, \end{cases}$$

i.e. we may write

$$\omega\big((a_{ji} u^* p_{li} u)^\natural\big) = \omega\big((q_{ji} p_{li})^\natural\big) \quad \text{for all } j, l\,.$$

The latter, together with the definition

$$a_i = \sum_j \beta_j a_{ji}\,, \tag{32}$$

leads to $\omega((a_i u^* p_{li} u)^\natural) = \omega((a p_{li})^\natural)$, and $(a_i u^* p_{li} u)^\natural \geqq (a_i v p_{li} v^*)^\natural$ for all $v \in \mathcal{U}$, where (31) has been inserted.

By (27), and taking into account the choice of u, we continue with the relations

$$\omega(z_i(a_i p_l)^\natural) = \omega((a p_{li})^\natural) \geqq \omega((a v p_{li} v^*)^\natural)\,, \tag{33}$$

$$(a_i p_l)^\natural z_i \geqq (a_i v p_l v^*)^\natural z_i \quad \text{for all } v \in \mathcal{U}\,. \tag{34}$$

From (33) we get $\omega(z_i(a_i p_l)^\natural) = K(\omega((a.)^\natural), p_l z_i)$ for all l.

By the same procedure, we may assure ourselves of a_i to every index i. Thus we may sum up over (33) with the result

$$\sum_i \omega(z_i(a_i p_l)^\natural) = K(\omega((a.)^\natural), p_l) \quad \text{for all } l,$$

and (iii) is seen; (ii) is given in the form of (34), for every i, and (30) and (32), finally, prove (i) to be true. ∎

Lemma 6-12. *Let $\mathcal{N}$ be a finite W^*-algebra, $o < p_1 < \cdots < p_n = 1$ projections. Suppose ϱ to be a normal positive linear functional on $\mathcal{N}$. Then, there is a $\{p_i\}$-canonical $\omega \in \mathcal{N}_+^*$ such that $\omega(p_i) = K(\omega, p_i) = K(\varrho, p_i)$ for all i.*

Proof. Let $\{\tau_k\}$ be a sequence of normal tracial states over $\mathcal{N}$, and $\{b_k\} \subset \mathcal{N}_+$, such that

$$\|\varrho - \tau_k(b_k\cdot)\|_1 \leqq k^{-1} \quad \text{for all } k = 1, 2, 3, \ldots \tag{34}$$

As usual, we may assume $b_k \in \mathcal{N}_+^0$ for all k. Fix k.

Application of Lemma 6-11 under the special choice $\omega = \tau_k$, $a = b_k$ provides $\{z_i\}$ and $\{a_i\}$, obeying (i)—(iii) of Lemma 6-11. We define (τ_k is tracial!)

$$\omega_k = \sum_i \tau_k(z_i(a_i\cdot)^\natural) = \sum_i \tau_k(z_i a_i\cdot)\,.$$

By Lemma 6-11 (ii) and (iii) we then have

$$K(\omega_k, p_l) = \omega_k(p_l) = K(\tau_k(b_k\cdot), p_l) \quad \text{for all } l\,. \tag{35}$$

By Lemma 6-11 (i) we are sure that

$$(a_i\cdot)^\natural z_i = \sum_s \beta'_{is}(p_s\cdot)^\natural z_i = \sum_s \beta'_{is}(p_s\cdot p_s)^\natural z_i\,, \tag{36}$$

with certain reals $\beta'_{is} \geqq 0$. It is easy to recognize that the restriction of $\tau_k(z_i(p_s\cdot)) = \tau_k(z_i(p_s\cdot p_s)^\natural)$ to $p_s \mathcal{N} p_s$ is tracial there.

Thus, and by the construction, the functional itself has to be a multiple of a certain state $\omega_{iks} \in \mathcal{S}_\infty^{p_s}$. Hence from (36) we come to

$$\tau_k(z_i(a_i\cdot)^\natural) = \sum_s \gamma_{is}\omega_{iks} \quad \text{with} \quad \omega_{iks} \in \mathcal{S}_\infty^{p_s}\,, \tag{37}$$

and $\gamma_{is} \geqq 0$ reals. With a view to (20′), (20″), we see that $\delta_{ks}\omega_{ks} = \sum_i \gamma_{is}\omega_{iks}$, with certain $\omega_{ks} \in \mathcal{S}_\infty^{p_s}$, $\delta_{ks} \geqq 0$ reals. Hence, because of (37) and due to the definition of ω_k, we arrive at

$$\omega_k = \sum_{s=1}^{n} \delta_{ks}\omega_{ks}\,, \qquad \omega_{ks} \in \mathcal{S}_\infty^{p_s}\,, \quad \delta_{ks} \geqq 0\,. \tag{38}$$

By (34) and (35) it follows that $\{\omega_k\}$ is a bounded set (it is $\|\omega_k\|_1 \leqq 1 + \|\varrho\|_1$ for every k). Then, since the number n does not depend on k, we are sure to find a subset $\left\{\omega_{k_\lambda} = \sum_{s=1}^{n} \delta_{k_\lambda s}\omega_{k_\lambda s}\right\}_{\lambda \in \Lambda}$ of $\{\omega_k\}$ such that $\lim_\lambda \delta_{k_\lambda s} = \delta_s \geqq 0$, $w^*\text{-}\lim_\lambda \omega_{k_\lambda s} = \omega_s'$, $w^*\text{-}\lim_\lambda \omega_{k_\lambda} = \omega$. Since $\mathcal{S}_\infty^{p_s}$ is w^*-closed (this is a consequence of the w^*-compactness of $\mathcal{S}_\infty(p_s\mathcal{N}p_s)$), we have $\omega_s' \in \mathcal{S}_\infty^{p_s}$, and $\omega = \sum_{s=1}^{n} \delta_s\omega_s'$, $\omega_s' \in \mathcal{S}_\infty^{p_s}$, $\delta_s \geqq 0$.

On the other hand, from (35) together with (34) (since $\tau_{k_\lambda}(b_{k_\lambda}\cdot) \to \varrho$ in norm) and $w^*\text{-}\lim_\lambda \omega_{k_\lambda} = \omega$, we arrive at

$$\omega(p_l) = K(\varrho, p_l) \quad \text{for all } l\,, \tag{39}$$

where we used the fact that K-functionals are continuously (in norm) depending on both arguments. Moreover, since ω is $\{p_s\}$-canonical and $\mathcal{N}$ is finite, Lemma 6-9 applies with $w = 1$, i.e. $K(\omega, p_l) = \sum_s r_s K(\omega_s, p_l) = \omega(p_l)$. The latter, however, together with (39) provides the asserted equality. ∎

Now we are ready to prove the "finite" variant of Proposition 6-10.

Proposition 6-13. *Let $\mathcal{M}$ be a finite W^*-algebra, $\varrho \in \mathcal{S}$. Suppose $o < p_1 < \cdots < p_n = 1$ to be projections of $\mathcal{M}$. There is a $\{p_i\}$-canonical positive linear form ω on $\mathcal{M}$ such that $\omega(p_i) = K(\omega, p_i) = K(\varrho, p_i)$ for all i.*

Proof. Throughout, we shall think of $\mathcal{M}$ as canonically embedded into its second dual $\mathcal{M}^{**}$. By $\mathcal{S}^{**}$ the state space of $\mathcal{M}^{**}$ will be meant, and $\mathcal{S}$ will be identified in the usual manner with the normal states of $\mathcal{M}^{**}$. K^{**} marks the Ky Fan functionals with respect to $\mathcal{M}^{**}$. Assume $\nu \in (\mathcal{M}^{**})_+^*$, $a \in \mathcal{M}_+^{**}$. Then, $K(\nu, a)$ will be understood as $K(\nu, a) = \sup_{u \in \mathcal{U}} \nu^u(a)$, with $\mathcal{U}$ being the unitaries of $\mathcal{M}$. Especially, in the case that $a \in \mathcal{M}$ we have $K(\nu, a) = K(\nu|_\mathcal{M}, a)$, the ordinary Ky Fan functional of the ν corresponding element of $\mathcal{M}_+^*$.*)

In the first step of our proof we are going to replace $\varrho \in \mathcal{S}$ by another state $\nu \in \mathcal{S}$ by the following procedure: Because of the Σ-property, we find unitaries $\{u_k\} \subset \mathcal{U}$ satisfying $\lim_k \varrho^{u_k}(p_i) = K(\varrho, p_i)$ for all i. Let $\{\varrho_{k_\lambda}\}_{\lambda \in \Lambda}$ be a universal subnet of $\{\varrho_k = \varrho^{u_k}\}_k$. Since $\mathcal{S}$ is w^*-compact $\{\varrho_{k_\lambda}\}$ converges weakly towards a state $\nu \in \mathcal{S}$. Because of $\varrho_k \sim \varrho$ ($\sim$ in sense of $(\mathcal{S}, \succ)$) we have $\varrho_k \succ \varrho$ for all k, i.e. $\varrho_{k_\lambda} \succ \varrho$. Therefore, and since the set of all states which are more mixed than a given ϱ is w^*-compact, we have $\nu = w^*\text{-}\lim_\lambda \varrho_{k_\lambda} \succ \varrho$, i.e. especially $K(\nu, p_i) \leqq K(\varrho, p_i)$ for all i. By definition of u_k, however,

$$\lim_\lambda \varrho_{k_\lambda}(p_i) = \nu(p_i) = K(\varrho, p_i) \geqq K(\nu, p_i) \geqq \nu(p_i) \quad \text{for all } i\,.$$

Hence equality has to occur:

$$\nu \in \mathcal{S}\,, \quad \nu \succ \varrho \quad \text{with} \quad \nu(p_i) = K(\nu, p_i) = K(\varrho, p_i) \quad \text{for all } i\,. \tag{40}$$

Let us note another useful relation:

---

*) Compare this to Chapter 5, (4).

Let z be a central projection of $\mathcal{M}^{**}$, and $a \in \mathcal{M}_+$. Then, under the assumption that $\sigma(a) = K(\sigma, a)$, $\sigma \in \mathcal{S}$, we have

$$\sigma(za) = K^{**}(\sigma, az) = K(\sigma, az) . \tag{41}$$

In fact, because of Chapter 5, (7), we see $K(\sigma, az) = K(\sigma(z.), a) = K^{**}(\sigma(z.), a) = K^{**}(\sigma, az) \geqq \sigma(az)$, and analogously, $K(\sigma, az^\perp) = K^{**}(\sigma, az^\perp) \geqq \sigma(az^\perp)$ $(\sigma(z.)$ resp. $\sigma(z^\perp.)$ are understood as normal elements of $\mathcal{S}^{**})$.

By assumptions on σ, a, and respecting Chapter 5, (7), we then conclude

$$\sigma(a) = K(\sigma, a) = K^{**}(\sigma, a) = K^{**}(\sigma, az) + K^{**}(\sigma, az^\perp)$$
$$= K(\sigma, az) + K(\sigma, az^\perp) \geqq \sigma(az) + \sigma(az^\perp) = \sigma(a) ,$$

i.e. equality is is required, and (41) has to be true.

Let $\{z_j\}$ be a finite orthogonal decomposition of 1 into central projections of $\mathcal{M}^{**}$ such that $z_1\mathcal{M}^{**}$ is the finite part of $\mathcal{M}^{**}$, and for $j \neq 1$ the projection sets $\{z_jp_i\}_i \subset z_j\mathcal{M}^{**}$ obey conditions (a)—(d) of Lemma 6-4, with respect to $\mathcal{M}^{**}z_j$.

Fix j. Then, we rename $\{z_jp_i\}_i \subset z_j\mathcal{M}^{**}$ into

$$(p_{01} < \cdots < p_{0n_0}) < (p_{11} < \cdots < p_{1n_1} < \cdots < p_{m^1} < \cdots < p_{mn_m} = z_j) , \tag{42}$$

in order to successfully adapt an application of Lemma 6-4. The parentheses indicate that the absence of the finite projections (p_{0i}) can also happen (this is possible in case of $j \neq 1$, $z_1 \neq 1$), or all the projections are finite (i. e. $j = 1$, $z_1 \neq o$). In the following we will concentrate our work on the "mixed" situation. (The remaining cases can be derived without problems). Let us look at the finite W^*-algebra $p_{0n_0}\mathcal{M}^{**}p_{0n_0} = \mathcal{N}$. Since ν from (40) in consideration over $\mathcal{M}^{**}$ is normal, $\nu|_{\mathcal{N}}$ is normal over $\mathcal{N}$, too. By Lemma 6-12 there is a $\{p_{0i}\}$-canonical $\nu_0^j \in \mathcal{N}_+^*$, with

$$\nu_0^j(p_{0i}) = K_{\mathcal{N}}(\nu_0^j, p_{0i}) = K_{\mathcal{N}}(\nu, p_{0i}) = \nu(p_{0i}) = K(\nu, p_{0i}) \quad \text{for all } i , \tag{43}$$

where $\nu_0^j = \sum_i \alpha_i\nu_i$, $\alpha_i \geqq 0$ reals, and $\nu_i \in \mathcal{S}_\infty^{p_{0i}}(\mathcal{N})$ for all i.

To see the last part in (43) one has to make reference to Lemma 6-6 (note that $\mathcal{N} \subset \mathcal{M}^{**}z_j$), (40) and (41) (with $a = p_i$, $z = z_j$). By definition of $\mathcal{S}_\infty^{p_{0i}}(\mathcal{N})$, and because of $\mathcal{N} \subset \mathcal{M}^{**}$, this set may be identified in a natural manner with a subset of $\mathcal{S}_\infty^{**p_{0i}}$. Hence $\sum_i \alpha_i\nu_i$ makes sense on $\mathcal{M}^{**}$, too. Having this in mind, we shall define

$$\nu^j = \sum_i \alpha_i\nu_i \in (\mathcal{M}^{**})_+^* , \qquad \nu_i \in \mathcal{S}_\infty^{**p_{0i}} , \tag{44}$$

and we may be sure of $\nu^j|_{\mathcal{N}} = \nu_0^j$.

Since $\mathcal{N}$ is finite, Lemma 6-9 applies with $w = 1$.

$$K_{\mathcal{N}}(\nu_s, p_{0i}) = \nu_s(p_{0i}) \quad \text{for all } i . \tag{45}$$

The special choice of ν_s implies that for $i \geqq s$ the following happens:

$$1 = \nu_s(p_{0s}) = K_{\mathcal{N}}(\nu_s, p_{0i}) \leqq K(\nu_s, p_{0i}) \leqq K^{**}(\nu_s, p_{0i}) \leqq \nu_s(1) = 1 , \tag{46}$$

i.e. equality has to occur.

In case of $i \leqq s$, Lemma 6-6 becomes applicable with the result

$$K_{\mathcal{N}}(\nu_s, p_{0i}) = K^{**}(\nu_s, p_{0i}) \geqq K(\nu_s, p_{0i}) \geqq \nu_s(p_{0i}) = K_{\mathcal{N}}(\nu_s, p_{0i}) , \tag{47}$$

where (45) has been inserted. From (45), (46) and (47) we arrive at

$$K_{\mathcal{N}}(\nu_s, p_{0i}) = K(\nu_s, p_{0i}) = K^{**}(\nu_s, p_{0i}) = \nu_s(p_{0i}) \quad \text{for all } i\,. \tag{48}$$

Taking into account the latter, by means of Lemma 6-9 from (43) and (44) follows that

$$\nu^j(p_{0i}) = K(\nu^j, p_{0i}) = K^{**}(\nu^j, p_{0i}) = \nu(p_{0i}) \quad \text{for all } i. \tag{49}$$

In the next steps we shall deal with the properly infinite projections, i.e. we consider

$$p_{11} < \cdots < p_{1n_1} < p_{21} < \cdots < p_{2n_2} < \cdots < p_{n1} < \cdots < p_{mn_m} = z_j\,. \tag{50}$$

By assumptions on (50), we have $p_{s1} \sim p_{sn_s}$ in $\mathcal{M}^{**}$ for every $s \geqq 1$, i.e.

$$K^{**}(\nu, p_{s1}) = K^{**}(\nu, p_{sn_s}) \quad \text{for all } s \geqq 1\,, \tag{51}$$

by Theorem 4-6.

Let $p_{si} = p_k z_j$ for an index k. Then, with a view to (40) and (41) (take $a = p_k$ there), we obtain

$$K(\nu, p_{si}) = K^{**}(\nu, p_{si})\,. \tag{52}$$

Because of (51), and applying (41) successively, from (52) we may derive the following relation:

$$\nu(p_{s1}) = \nu(p_{s2}) = \cdots = \nu(p_{sn_s}) \quad \text{for all } s \geqq 1\,. \tag{53}$$

Let us choose $\{\sigma_s\}_{s=1}^m \subset \mathcal{S}^{**}$ and reals $\{t_s\}_{s=1}^m$ such that

$$\sigma_s \in \mathcal{S}_\infty^{**p_{s1}}\,, \quad \text{and} \quad t_s = \nu(p_{s1}) - \nu(p_{s-1,\,n_{s-1}}) \quad \text{for all } s \geqq 1\,. \tag{54}$$

With these settings, we define states σ^j by:

$$\sigma^j = \sum_s t_s \sigma_s\,. \tag{55}$$

Let $\mathcal{J}_p^{**}$ denote the c-ideal in $p\mathcal{M}^{**}p$, with $p \in \mathcal{M}^{**p}$. Then, due to (50), (53), (54), (55) and Lemma 6-6 — and since $p_{s1} \in \mathcal{J}^{**}_{p_{s+1,1}}$ for all s by assumptions on the projections in (50) — it is easy to check that

$$\begin{aligned} K^{**}(\sigma^j, p_{si}) = \sigma^j(p_{si}) &= \sum_{1 \leqq k \leqq s} t_k \sigma_k(p_{s1}) = \sum_{1 \leqq k \leqq s} t_k \\ &= \nu(p_{s1}) - \nu(p_{0n_0}) = \nu(p_{si}) - \nu(p_{0n_0})\,. \end{aligned} \tag{56}$$

Since p_{0n_0} is finite, we have $p_{0n_0} \in \mathcal{J}^{**}_{p_{11}}$, i.e.

$$K^{**}(\sigma^j, p_{0i}) = 0 \quad \text{for all } i\,. \tag{57}$$

Since $\mathbf{1} \in \mathcal{U} \subset \mathcal{U}^{**}$, (56) and (57) result in:

$$K(\sigma^j, p_{si}) = \nu(p_{si}) - \nu(p_{0n_0}) \quad \text{for all } s \geqq 1\,, \tag{58}$$

and

$$K(\sigma^j, p_{0i}) = 0 \quad \text{for all } i\,. \tag{59}$$

Let $k_1 < \cdots < k_m \leqq n$ be chosen in such a way that

$$p_{s1} = p_{k_s} z_j\,, \quad \text{and for} \quad l < k_s\colon\ p_l z_j \neq p_{s1}\,. \tag{60}$$

Remembering the fact that we started with a finite W^*-algebra $\mathcal{M}$, we see that the original $p_i\mathcal{M}p_i$ is finite, too. Let $\natural_i$ denote the operation "centre-valued trace" with respect to $p_i\mathcal{M}p_i$. Then, the induced operation in the embedding of $p_i\mathcal{M}p_i$ into $\mathcal{M}^{**}$ will be denoted by the same sign (this makes sense since $\{a^{\natural p}\} = ||$-cl conv $(a^u)_{u\in\mathcal{U}_p} \cap \mathcal{Z}p$, and the embedding of $p\mathcal{M}p$ into $\mathcal{M}^{**}$ via the embedding of $\mathcal{M}$ into $\mathcal{M}^{**}$ is isometric). Then, we consider

$$\sigma'_s = \sigma_s(z_j(p_{k_s}\cdot p_{k_s})^{\natural k_s})|_{\mathcal{M}} \in \mathcal{S} . \tag{61}$$

Assume $l < k_s$. Then, by Lemma 6-6:

$$K(\sigma'_s, p_l) = K_{p_{k_s}}(\sigma'_s, p_l) = K_{p_{k_s}}(\sigma'_s(z_j(.)^{\natural k_s}), p_l) = \sigma_s(z_j p_l^{\natural k_s}) . \tag{62}$$

Now, $p_l^{\natural k_s} \in ||$-cl conv $\{vp_lv^*\}_{v\in\mathcal{U}_{p_{k_s}}}$, with $\mathcal{U}_{p_{k_s}}$ the unitaries in $p_{k_s}\mathcal{M}p_{k_s}$. But then, also $p_l^{\natural k_s}z_j \in ||$-cl conv $\{vp_lz_jv^*\}_{v\in\mathcal{U}_{p_{k_s}}}$ with respect to $\mathcal{M}^{**}$ (canonical imbedding is isometric). Since c-ideals are uniformly closed ones, and since $p_lz_j \in \mathcal{J}^{**}_{p_{s1}}$ by (60) and our assumptions, we have $p_l^{\natural k_s}z_j \in \mathcal{J}^{**}_{p_{s1}} = \mathcal{J}^{**}_{p_{k_s}}z_j$. From this argumentation, with a view to (62) and the definition of σ_s we derive

$$K(\sigma'_s, p_l) = 0 \quad \text{in the case of } l < k_s . \tag{63}$$

In the case of $l \geqq k_s$ it is obvious that

$$K(\sigma'_s, p_l) = \sigma_s(p_{s1}) . \tag{64}$$

By definition of σ'_s, $\sigma'_s|_{p_{k_s}\mathcal{M}p_{k_s}}$ is a tracial state. Thus

$$\sigma'_s|_{p_{k_s}\mathcal{M}p_{k_s}} \in \mathcal{S}_\infty(p_{k_s}\mathcal{M}p_{k_s}) . \tag{65}$$

We define another functional on $\mathcal{M}$ by setting

$$\sigma'^j = \sum_{s\geqq 1} t_s\sigma''_s , \quad \text{with} \quad \sigma''_s = \sigma'_s(p_{k_s}\cdot p_{k_s}) \in \mathcal{S}^{p_{k_s}}_\infty . \tag{66}$$

With this definition and by (63), (64) and (65) we arrive at

$$K(\sigma'^j, p_l) = \begin{cases} 0 \quad \text{for } l < k_1 , \\ \sum_{1\leqq k\leqq s} t_k\sigma_k(p_{s1}) = \sum_{1\leqq k\leqq s} t_k\sigma_k(p_lz_j) = \sum_{1\leqq k\leqq s} t_k = \nu(p_lz_j) - \nu(p_{0n_0}) \\ \text{in case of } k_s \leqq l < k_{s+1} , \end{cases} \tag{67}$$

where by definition $k_{m+1} = n$ if $k_m \neq n$, and (53) has been used (in case of $k_s \leqq l < k_{s+1}$ one has $\nu(p_lz_j) = \nu(p_{k_s}z_j)$). The index s may range from 1 to m, so (67), (63), (64) together imply

$$K(\sigma'^j, p_l) = \left\{\begin{matrix} 0 \quad \text{for } l < k_1 \\ \nu(p_lz_j) - \nu(p_{0n_0}) \quad \text{for } l \geqq k_1 \end{matrix}\right\} = \sigma'^j(p_l) . \tag{68}$$

There exist indices $l_1 < l_2 < \cdots < l_{n_0} < k_1$ with

$$p_{0i} = p_{l_i}z_j , \quad \text{and} \quad p_lz_j \in \{p_{0i}\}_i \quad \text{for every } l < k_1 . \tag{69}$$

Because of (44), i.e. since $\nu_i \in \mathcal{S}^{**p_{0i}}_\infty$ and p_{0i} is finite in $\mathcal{M}^{**}z_j$, $\nu_i(z_j\cdot)|_{p_{l_i}\mathcal{M}p_{l_i}}$ is a tracial state, so

$$\nu'_i = \nu_i(z_j\cdot)|_{p_{l_i}\mathcal{M}p_{l_i}} \in \mathcal{S}_\infty(p_{l_i}\mathcal{M}p_{l_i}) . \tag{70}$$

Together with (44) we see

$$\nu^j|_{\mathcal{M}} = \sum_i \alpha_i\nu''_i = \nu'^j \quad \text{with} \quad \nu''_i = \nu'_i(p_{l_i}\cdot p_{l_i}) \in \mathcal{S}^{p_{l_i}}_\infty . \tag{71}$$

For $p_l z_j = p_{0i}$ (i.e. $l < k_1$) we get by means of (49) and (71)

$$\nu'^j(p_l) = \nu^l(p_l z_j) = K(\nu^j, p_l z_j) = K(\nu'^j, p_l) = \nu(p_l z_j) = \nu(p_{0i}) , \tag{72}$$

and for $l \geqq k_1$:

$$\nu'^j(p_l) = \nu'^j(p_{l_{n_0}}) = \nu^j(p_{l_{n_0}} z_j) = \nu^j(p_{0n_0}) = \nu(p_{0n_0}) = K(\nu'^j, p_l) . \tag{73}$$

Finally, if we define ω^j to be

$$\omega^j = \nu'^j + \sigma'^j \quad \text{on } \mathcal{M} , \tag{74}$$

we see from (68), (72) and (73) that

$$\omega^j(p_l) = K(\omega^j, p_l) = \nu(p_l z_j) \quad \text{for all } l . \tag{75}$$

By the remark following (42) and the constructions above we arrive at functionals of type (74) also in the situation where only finite or only properly finite projections occur. (Then we have either ν'^j or σ'^j vanishing.)

Let us consider the whole set of these ω^j functionals. Then, by construction, both, ν'^j and σ'^j, are $\{p_l\}$-canonical functionals (see (66), (71)). Hence, by (20), $\omega = \sum_j \omega^j$ appears to be $\{p_l\}$-canonical. Because of (75) we have

$$\begin{aligned} \nu(p_l) &= \sum_l \nu(p_l z_j) = \sum_j \omega^j(p_l) = \omega(p_l) = \sum_j K(\omega^j, p_l) \\ &\geqq K\Big(\sum_j \omega^j, p_l\Big) \geqq \sum_j \omega^j(p_l) = \omega(p_l) . \end{aligned}$$

Equality occurs and we may write

$$\nu(p_l) = \omega(p_l) = K(\omega, p_l) \quad \text{for all } l. \tag{76}$$

This, however, gives, because of (40),

$$\omega(p_l) = K(\omega, p_l) = K(\varrho, p_l) \quad \text{for all } l . \tag{77}$$

with $\{p_l\}$-canonical ω, and the proof is complete. ∎

In proving Proposition 6-13 we have shown the Proposition 6-10 to be a true one for every finite W^*-algebra.

The general case now can be treated with the help of a modification of the previous proof's construction scheme. One can do this in a manner which completely avoids referring to the second dual by replacing all manipulations by operations within the algebra $\mathcal{M}$ itself. Indeed, Proposition 6-13 can serve as an appropriate tool for avoiding normality requirements in the derivation of an equivalent to (43). Then we are allowed to work in $\mathcal{M}$ exclusively, and, taking the proof above as a guide, we can follow the line of reasoning indicated by (44), (49), (54), (55), (58), (59), (74), (76), (77). In this way it is possible to prove the Assertion 6-10 in the general situation, too. These remarks should be sufficient to see how a complete proof could run and we omit a detailed exposition.

6.5. A characterization of $(\mathcal{M}_+, \succ)$

As indicated in the introduction, we are going to formulate and prove a certain "minimal" functional characterization of $(\mathcal{M}_+, \succ)$, see [118, 13].

Theorem 6-13. *Let $\mathcal{M}$ be a W^*-algebra. Then, for $a, b \in \mathcal{M}_+$ we have $a \succ b$ if and only if $K(\omega, a) \geqq K(\omega, b)$ for each $\omega \in \mathcal{S}^p_\infty$ and for all $p \in \mathcal{M}^p$.*

Proof. Necessity of the formulated conditions is obvious due to the definition of $\succ$.

Sufficiency: By Remark 6-2, we may be content with showing the validity of the implication:

$$K(\varrho, a') \geqq K(\varrho, b') \quad \text{for all } \varrho \in \mathcal{S}^p_\infty \text{ and for all } p \in \mathcal{M}^p \text{ implies } a' \succ b' \,, \tag{78}$$

for every $a' \in U(a)$ and each $b' \in L(b)$. Let $a' \in U(a)$, $b' \in L(b)$, and consider them fixed in the sequel.

By standard arguments we find a finite orthogonal decomposition of $\mathbf{1}$ into central projections $\{z_i\}$ such that $a_i = a'z_i$, $b_i = b'z_i$ for each i are of the type announced in Remark 6-3.

Fix i. Then,

$$a_i = \sum_j \alpha_j p_j \,, \qquad b_i = \sum_l \beta_l q_l \,, \tag{79}$$

with reals $\alpha_j, \beta_l \geqq 0$ and $\{p_i\}$, $\{q_l\}$ obeying conditions (1)—(3) (see Remark 6-3), with $p_n = q_m = z_i$. By Lemma 6-4 partial isometries are provided, V and Y, with

$$b_i' = Vb_iV \geqq b_i \,, \qquad a_i' = Ya_iY^* \geqq a_i \,, \qquad VV^* = YY^* = z_i \,, \tag{80}$$

and the b_i', a_i' corresponding ordered sequences of projections obey assumptions for an application of Lemma 6-5, in case of the W^*-algebra $z_i\mathcal{M}$. Applying Lemma 6-5, we find a unitary $u \in z_i\mathcal{M}$ such that the union of those ordered sets of projections which correspond to $b_i'' = ub_i'u^*$ and a_i', respectively, give an $\leqq$-ordered projection system $\{Q_k\}$.

With certain non-negative reals t_k, t_k' we may write

$$b_i'' = \sum_k t_k Q_k \,, \qquad a_i' = \sum_k t_k' Q_k \,. \tag{81}$$

Because of (80), $b_i'' = uVa_iV^*u^*$ and Lemma 3-4, we have

$$K(\omega, b_i) = K(\omega, b_i'') \,, \qquad K(\omega, a_i) = K(\omega, a_i') \quad \text{for all } \omega \in \mathcal{M}_+^* \,. \tag{82}$$

We notice that, up to now, the proof followed the same line as in 6.3.

Assume $\varrho \in \mathcal{M}_+^*$ is fixed. Then, Proposition 6-10 provides a $\{Q_k\}$-canonical $\omega \in \mathcal{M}_+^*$ such that

$$\omega(Q_k) = K(\omega, Q_k) = K(\varrho, Q_k) \quad \text{for all } k \,. \tag{83}$$

By virtue of the Σ-property, from (82), (83) together with (81) it follows that

$$K(\varrho, b_i) = \sum_k t_k K(\varrho, Q_k) = \sum_k t_k K(\omega, Q_k) = K(\omega, b_i'') = \omega(b_i'') \,, \tag{84}$$

and, analogously,

$$K(\varrho, a_i) = K(\omega, a_i') = \omega(a') \,. \tag{85}$$

Assume $\omega = \sum_k \delta_k \omega_k$, with reals $\delta_k \geqq 0$ and $\omega_k \in \mathcal{S}^{Q_k}_\infty$. Now, because of $\mathcal{S}^{pz}_\infty \subset \mathcal{S}^p_\infty$ for any central projection $z \neq o$, from the assumptions of (78) it obviously follows that

$$K(\nu, a_i) \geqq K(\nu, b_i) \quad \text{for all } \nu \in \mathcal{S}^p_\infty \text{ and for all } p \in \mathcal{M}^p \,. \tag{86}$$

Taking into account (86) and (80) one gets

$$K(\omega_k, b_i'') = K(\omega_k, b_i) \leqq K(\omega_k, a_i) = K(\omega_k, a_i'), \tag{87}$$

for each k. Since ω is $\{Q_k\}$-canonical, Lemma 6-9 applies with the result that

$$K(\omega, Q_j) = \sum_k \delta_k K(\omega_k, Q_j) \quad \text{for all } j. \tag{88}$$

Respecting (81) and (88), and due to the Σ-property we may conclude that

$$K(\omega, b_i'') = \sum_k \delta_k K(\omega_k, b_i'') \quad \text{and} \quad K(\omega, a_i') = \sum_k \delta_k K(\omega_k, a_i'). \tag{89}$$

Due to (87), and since $\delta_k \geqq 0$ for every k, we see $K(\omega, b_i'') \leqq K(\omega, a_i')$. In view of (84) and (85), the last inequality enables us to write

$$K(\varrho, b_i) \leqq K(\varrho, a_i). \tag{90}$$

Inequality (90) may be derived for every index i, and since $b' = \sum_i b_i$ and $a' = \sum_i a_i$, with $\{z_i\} \subset \mathcal{Z}^p$ and $b_i = b'z_i$, $a_i = a'z_i$, we may take sums

$$K(\varrho, b') = \sum_i K(\varrho, b_i) \leqq \sum_i K(\varrho, a_i) = K(\varrho, a').$$

Since ϱ could have been chosen arbitrarily from $\mathcal{M}_+^*$, the relation $a' \succ b'$ is valid and (78) is established. Since $a' \in U(a)$, $b' \in L(b)$, but arbitrarily, (78) has to be valid for any such pair a', b', so we have $a \succ b$ by Remark 6-2. ▮

6.6. The main theorem

In this section we aim at verifying Theorem 6-1. Whenever $Q \in \mathcal{M}^p$, the §-operation on $Q\mathcal{M}Q$ will be marked as $\S_Q$. We start with (cf. [118]):

Lemma 6-14. *Let $\mathcal{M}$ be a W^*-algebra, and Q be a projection of $\mathcal{M}$. Let $a \in \mathcal{M}_+$. There is a unique $\leqq$-maximal element, $a^{\&Q}$ in sign, in the uniform closure $\{(Qbab^*Q)^{\S Q}\}_{bb^*=1}$.*

Proof. Let $o \leqq a_1 \leqq \cdots \leqq a_n$ be elements of $\mathcal{M}$, with $a_n \in L(a)$ for all n, and $\|\text{-}\lim_n a_n = a$. Fix n, and suppose a_n to be represented as

$$a_n = \sum_i \beta_i p_i, \qquad \beta_i \geqq 0 \quad \text{for all } i, \quad o = p_0 < p_1 < \cdots < p_m = 1, \tag{91}$$

with (p_i) projections. Arguing as in 6.2., we find a finite orthogonal decomposition of $\mathbf{1}$ into central projections $\{z_j\}$ such that, for any index j, the systems $\{p_i z_j\}_i$ and $\{Qz_j\}$ obey conditions described in (1)—(3). Then, in applying Lemma 6-4 with respect to both the projection sets, we are provided with a partial isometry v_j, $v_j v_j^* = z_j$, such that $\{v_j p_i v_j^* z_j\}_i$, $\{Qz_j\}$ obey all the assumptions for an application of Lemma 6-5.

The latter then guarantees unitary elements $u_j \in \mathcal{M}z_j$ with:

$$\{u_j v_j p_j v_j^* u_j^*, Qz_j\}_i \quad \text{is } \geqq\text{-ordered.}$$

Defining $w_n = \sum_j u_j v_j$ it is plain to see the relations $w_n w_n^* = 1$ and

$$\{w_n p_i w_n^* z_j, Qz_j\}_i \quad \text{is } \geqq\text{-ordered for any } j. \tag{92}$$

Let $\omega \in \mathcal{S}_\infty^Q$. Then

$$K(\omega, w_n p_i w_n^*) = \sum_j K(\omega, w_n p_i w_n^* z_j) . \tag{93}$$

Fix j. Then, because of (92), we have

$$\text{either} \quad Qz_j > w_n p_i w_n^* z_j , \qquad \text{or} \quad w_n p_i w_n^* z_j \geqq Qz_j . \tag{94}$$

If the first part in (94) happens, we may use Lemma 6-6 to draw the conclusion that

$$K(\omega, w_n p_i w_n^* z_j) = K_{Qz_j}(\omega, w_n p_i w_n^* z_j) = K_{Qz_j}(\omega, w_n p_i w_n^*) . \tag{95}$$

This remains true for the second alternative in (94) because of the structure of ω (we also put: $K_p(\varrho, b) = \sup_{u \in \mathcal{U}_p} \varrho^u(b)$ with $\mathcal{U}_p$ the unitaries in $p\mathcal{M}p$, for $\varrho \in \mathcal{M}_+^*$ and $b \in \mathcal{M}$). Because of the Σ-property, (91), (94), and structure of ω, (91) implies

$$K(\omega, w_n a_n w_n^* z_j) = K_{Qz_j}(\omega, w_n a_n w_n^* z_j) .$$

Due to $\omega \in \mathcal{S}_\infty^Q$, $\omega(z_j \cdot)$ is a multiple of a state of $\mathcal{S}_\infty^{Qz_j}$. Hence Theorem 4-12 shows

$$K_{Qz_j}(\omega, w_n a_n w_n^* z_j) = K_{Qz_j}(\omega, Qw_n a_n w_n^* Qz_j) = \omega\big((Qw_n a_n w_n^* Qz_j)^{\S_{Qz_j}}\big) .$$

From the construction of $\S$-maps (Definition 4-1, (v)), and since $z_j \in \mathcal{Z}^p$, we have

$$(Qw_n a_n w_n^* Qz_j)^{\S_{Qz_j}} = z_j (Qw_n a_n w_n^* z_j)^{\S_Q} = z_j (Qw_n a_n w_n^* Q)^{\S_Q} .$$

Taking together these equalities, we arrive at

$$K(\omega, w_n a_n w_n^*) = \omega\big((Qw_n a_n w_n^* Q)^{\S_Q}\big) , \qquad w_n w_n^* = 1 . \tag{96}$$

By construction of $w_n z_j = u_j v_j$, with v_j and u_j chosen in accordance with Lemma 6-4 and Lemma 6-5, respectively, the following is valid: $w_n p_i w_n^* z_j \sim p_i z_j$ for all j, i.e. $w_n p_i w_n^* \sim p_i$. The latter, by Theorem 4-6, makes that $K(\omega, p_i) = K(\omega, w_n p_i w_n^*)$, so from (96) we see

$$K(\omega, a_n) = K(\omega, w_n a_n w_n^*) = \omega\big((Qw_n a_n w_n^* Q)^{\S_Q}\big) , \tag{97}$$

where the Σ-property has been used once again.

Let $b \in \mathcal{M}$, $bb^* = 1$. Then, because of $\|b\| \leqq 1$ and $\omega \in \mathcal{S}_\infty^Q$, one has

$$K(\omega, a_n) \geqq K(\omega, Qba_n b^* Q) = K_Q(\omega, Qba_n b^* Q) .$$

In the last step Lemma 3-4 has been used. By the structure of $\mathcal{S}_\infty^Q$ and Theorem 4-12, we have $\omega\big((Qba_n b^* Q)^{\S_Q}\big) = K_Q(\omega, Qba_n b^* Q)$. We may compose these inequalities ((97) included) to see a new one: $\omega\big((Qw_n a_n w_n^* Q)^{\S_Q}\big) \geqq \omega\big((Qba_n b^* Q)^{\S_Q}\big)$ for each b with $bb^* = 1$. Since $\omega \in \mathcal{S}_\infty^Q$, but taken at choice, the derived inequality is true for every $\omega \in \mathcal{S}_\infty^Q$. Now, $\S_Q$ is a map from $Q\mathcal{M}Q_+$ into $\mathcal{Z}_+Q$, and $\mathcal{S}_\infty^Q = \{\nu(Q \cdot Q)\}_{\nu \in \mathcal{S}_\infty(Q\mathcal{M}Q)}$ is satisfying $\mathcal{S}_\infty^Q|_{\mathcal{Z}Q} = \mathcal{S}(\mathcal{Z}Q)$ (see Chapter 4, (22), Theorem 4-11, and Theorem 4-12). Hence, the inequality derived, results in

$$(Qw_n a_n w_n^* Q)^{\S_Q} \geqq (Qba_n b^* Q)^{\S_Q} \quad \text{for all } b \in \mathcal{M},\ bb^* = 1 . \tag{98}$$

The scheme of construction previously demonstrated applies for each n, and inequality (98) is true for every n, with a partial isometry w_n, $w_n w_n^* = 1$. We are going to consider different indices n, m. Because of $a_n \geqq a_m$, in the case of $n \geqq m$, and (98) we obtain

$$(Qw_n a_n w_n^* Q)^{\S_Q} \geqq (Qw_m a_n w_m^* Q)^{\S_Q} \geqq (Qw_m a_m w_m^* Q)^{\S_Q}, \qquad n \geqq m. \tag{99}$$

From Definition 4-1, (iii), (iv), and Theorem 4-9 we get

$$o \leqq (Qw_n a_n w_n^* Q)^{\S_Q} - (Qw_n a_m w_n^* Q)^{\S_Q} \leqq \big(Qw_n(a_n - a_m)\, w_n^* Q\big)^{\S_Q} \quad \text{for all } n \geqq m,$$

so by Definition 4-1, (i), and Theorem 4-9

$$\|(Qw_n a_n w_n^* Q)^{\S_Q} - (Qw_n a_m w_n^* Q)^{\S_Q}\| \leqq \|a_n - a_m\| \quad \text{for all } n, m. \tag{100}$$

From (98) comes that $(Qw_m a_m w_m^* Q)^{\S_Q} \geqq (Qw_n a_m w_n^* Q)^{\S_Q}$, so (99) implies that

$$\begin{aligned} o &\leqq (Qw_n a_m w_n^* Q)^{\S_Q} - (Qw_m a_m w_m^* Q)^{\S_Q} \\ &= (Qw_n a_n w_n^* Q)^{\S_Q} - (Qw_n a_m w_n^* Q)^{\S_Q} + \{(Qw_n a_m w_n^* Q)^{\S_Q} - (Qw_m a_m w_m^* Q)^{\S_Q}\} \\ &\leqq (Qw_n a_n w_n^* Q)^{\S_Q} - (Qw_n a_m w_n^* Q)^{\S_Q}, \end{aligned}$$

whenever $n \geqq m$.

The latter, together with (100), implies that $\{(Qw_n a_n w_n^* Q)^{\S_Q}\}_n$ is a $\leqq$-increasing Cauchy sequence of elements in $\mathcal{Z}_+Q$. Hence, there is a limit element. Call this element $a^{\&_Q}$:

$$\|\text{-}\lim_n (Qw_n a_n w_n^* Q)^{\S_Q} = a^{\&_Q} \in \mathcal{Z}_+Q. \tag{101}$$

Then, due to (98) and (99) — and respecting the uniform continuity of $\S_Q$ — we are led to the conclusion

$$a^{\S_Q} \geqq (Qbab^*Q)^{\S_Q} \quad \text{for all } b, \quad bb^* = 1,$$

i.e. $a^{\&_Q}$ majorizes elements of the set in question.

Finally, because of (100), (101), and inserting the continuity argument once more, we have

$$a^{\&_Q} \in \|\text{-cl}\, \{(Qbab^*Q)^{\S_Q}\}_{bb^*=1}. \tag{102}$$

These results prove that $a^{\&_Q}$ is the asserted element. ∎

The family of $\&_Q$-operations has properties which remember some of the properties we have found in case of $\S$. One of them is in the following (compare this with Theorem 4-12):

Lemma 6-15. *Let $\mathcal{M}$ be a W^*-algebra, $a \in \mathcal{M}_+$ and $\omega \in \mathcal{S}_\infty^Q$, with $Q \in \mathcal{M}^p$ non-trivial. Then, $K(\omega, a) = \omega(a^{\&_Q})$.*

Proof. We choose $\{a_n\}$, $\{w_n\}$ as in the previous proof. Then, from (97), (98) and by definition of $\&_Q$ we conclude

$$K(\omega, a_n) = \omega\big((Qw_n a_n w_n^* Q)^{\S_Q}\big) = \omega(a_n^{\&_Q}) \quad \text{for all } n.$$

This, however, leads immediately to the asserted relation by (102) and uniform continuity of $\&_Q$ (which is a consequence of $\S_Q$-continuity, see also the proof of Lemma 4-7). ∎

Lemma 6-16. *Let $\mathcal{M}$ be a W^*-algebra, and $a, b \in \mathcal{M}_+$. Then $a \succ b$ if and only if $a^{\&Q} \geq b^{\&Q}$ for each non-zero $Q \in \mathcal{M}^Q$.*

Proof. By Theorem 6-13 and Lemma 6-15 $a \succ b$ iff

$$\omega(a^{\&Q}) \geq \omega(b^{\&Q}) \quad \text{for all } \omega \in \mathcal{S}^Q_\infty \text{ and for all } Q \in \mathcal{M}^P .$$

Because of $a^{\&Q}, b^{\&Q} \in \mathcal{Z}Q$, and since $\mathcal{S}^Q_\infty|_{\mathcal{Z}Q} = \mathcal{S}(\mathcal{Z}Q)$, we arrive at

$$a^{\&Q} \geq b^{\&Q} \quad \text{iff} \quad \omega(a^{\&Q}) \geq \omega(b^{\&Q}) \quad \text{for all } \omega \in \mathcal{S}^P_\infty .$$

Combining both statements the asserted equivalence can be seen; see [118]. ∎

It is very useful to notice the following properties of the operation $\&_Q$ (the non-trivial ones result from the Σ-property of states and arguments which we have repeatedly stressed):

Properties of $\&_Q$:

(a) $\|a^{\&Q}\| \leq \|a\|$ for all $a \in \mathcal{M}_+$;

(b) $1^{\&Q} = Q$ for all $Q \in \mathcal{M}^P$;

(c) $\&_Q$ is subadditive and monotone on $\mathcal{M}_+$;

(d) $\&_Q$ has the Σ-property, i.e. for projections $Q_1 < \cdots < Q_n$ and non-negative reals α_i:

$$\Big(\sum_i \alpha_i Q_i\Big)^{\&Q} = \sum_i \alpha_i Q_i^{\&Q} .$$

In knowing Lemma 6-16 and the above mentioned properties, we also have all the tools for a proof of Theorem 6-1.

Let $Q \in \mathcal{M}^P$, and assume $c(Q)$ is the central support of Q. Then, because of $a^{\&Q} \in \mathcal{Z}_+Q$ for all $a \in \mathcal{M}_+$, we have a uniquely determined map $\mathcal{J}_Q$ from $\mathcal{M}_+$ into $\mathcal{Z}_+$ such that

$$a^{\&Q} = \mathcal{J}_Q(a)\, Q\,, \qquad \mathcal{J}_Q(a) \in \mathcal{Z}c(Q) \quad \text{for all } a \in \mathcal{M}_+ . \tag{103}$$

Indeed, the decisive condition is the second one of (103). (This is easily recognized.) This condition implies

$$\left.\begin{aligned} &\|\mathcal{J}_Q(a)\, Q\| = \|\mathcal{J}_Q(a)\| \quad \text{for all } a \in \mathcal{M}_+\,, \\ \text{and} \\ &a^{\&Q} \leq b^{\&Q} \quad \text{for } a, b \in \mathcal{M}_+ \quad \text{iff} \quad \mathcal{J}_Q(a) \leq \mathcal{J}_Q(b)\,. \end{aligned}\right\} \tag{104}$$

Let $Q, P \in \mathcal{M}^P$, with $Q \sim P$. Then, there is a partial isometry w with $w^*w = Q$, $ww^* = P$, i.e. $w^*pw = Q$, $wQw^* = P$, and $w.w^*$ maps $Q\mathcal{M}Q$ isometrically onto $P\mathcal{M}P$. This gives by Theorem 4-12 that $\mathcal{S}^P_\infty = \{\varrho(w^*.w)\}_{\varrho \in \mathcal{S}^Q_\infty}$. Let $\varrho = \omega(w^*.w)$, for $\omega \in \mathcal{S}^Q_\infty$. Then, by Lemma 6-15 and Lemma 3-4

$$\varrho(a^{\&P}) = K(\varrho, a) = K\big(\omega(w^*.w), a\big) \leq K(\omega, a) = \omega(a^{\&Q}) \quad \text{for all } a \in \mathcal{M}_+ .$$

From this, and by definition of ϱ and since $Qa^{\&Q}Q = a^{\&Q}$, $ww^* = Q$ implies $w^*wa^{\&Q}w^*w = a^{\&Q}$, we obtain $\varrho(a^{\&P}) \leq \varrho(wa^{\&Q}w^*)$ for all $a \in \mathcal{M}_+$. All these conditions, together with $\mathcal{S}^P_\infty|_{\mathcal{Z}P} = \mathcal{S}(\mathcal{Z}P)$, guarantee the validity of $a^{\&P} \leq wa^{\&Q}w^*$. Since all the arguments apply after performing replacements $Q \to P$, $P \to Q$, $w \to w^*$, as well, we have $a^{\&Q} \leq w^*a^{\&P}w$. $ww^* = P$ and $a^{\&P} \in \mathcal{Z}P$ make that

$wa^{\&_Q}w^* \leq a^{\&_P}$. Taking together all this, we arrive at $a^{\&_P} = wa^{\&_Q}w^*$ for all $a \in \mathcal{M}_+$. Because of $ww^* = P$ and (103), we then see

$$\mathcal{J}_P(a) = \mathcal{J}_Q(a) \quad \text{in case of } Q \sim P,\ Q, P \in \mathcal{M}^p . \tag{104}$$

At last, again by definition of $\mathcal{J}_Q$ in (103), a Σ-property for $\mathcal{J}_Q$ follows from property (d) of $\&_Q$.

Finally, because of $a^*a \sim aa^*$ in sense of $(\mathcal{M}_+, \succ)$ (this follows from the corresponding behavior of Ky Fan functionals) we have from Lemma 6-16 and (103) validity of $\mathcal{J}_Q(a^*a) = \mathcal{J}_Q(aa^*)$, and we may formulate an extended version of Theorem 4-1:

Theorem 6-17 (see [118]). *Let $\mathcal{M}$ be a W^*-algebra. There exists a family $\{\mathcal{J}_q\}_{q \in \mathcal{M}^p}$ of maps from $\mathcal{M}_+$ into $\mathcal{Z}_+$, with the properties 1—9 of Theorem 6-1. In addition we have:*

(1′) $\quad \mathcal{J}_q(R) = \mathcal{J}_q(P) \quad$ *for* $\quad R \sim P, \quad R, P \in \mathcal{M}^p$;

(2′) $\quad \mathcal{J}_q(R)\, q = R^{\&_q} = R_0^{\&_q} \quad$ *if* $\quad R \sim R_0 \leq q, \quad R, R_0 \in \mathcal{M}^p$;

(3′) $\quad \mathcal{J}_q(R) = c(q) \quad$ *if* $\quad R \succ Q$;

(4′) $\quad \mathcal{J}_q = \mathcal{J}_q(z\,.) + \mathcal{J}_q(z^\perp .) = \mathcal{J}_{qz} + \mathcal{J}_{qz^\perp} \quad$ *for all* $\quad z \in \mathcal{Z}^p$.

Proof. Properties 1—9 of $\mathcal{J}_q$ as formulated in Theorem 6-1 follow from (103) and (104) in context with the discussion above. (1′) is a consequence of property **3**, (2′) is derived from the definition of $\mathcal{J}_q$ via (103) and the structure of $\&_q$. (3′) is true because of property (b) of $\&_q$, whereas (4′) is obvious from the definition of $\&_q$ and (103) again.

The properties of $\mathcal{J}_q$-maps just derived — especially the Σ-property — show the intrinsic characterization of $\{\mathcal{M}_+, \succ\}$ to be essentially a matter of knowing the centre-valued "matrix" $\{\mathcal{J}_q(p)\}_{q,p \in \mathcal{M}^p}$.

6.7. A special case: Finite W^*-algebras

In this section the $\succ$-relation on a finite W^*-algebra will be considered. In the case of finite W^*-algebras, the structure of $\mathcal{J}_q$-maps can be simplified. By means of this simplification we will be able to give another formulation of $\succ$ on $\mathcal{S}$, which is based on an extension to arbitrary finite W^*-algebras, of the notion of the $e_s(.)$-functionals as defined in 1.6., (27) (see also Theorem 2-2) for the type I_n factor M_n.

Let $\mathcal{M}$ be finite, and $Q \in \mathcal{M}^p$. As usual, $\natural$, $\natural_Q$ denote the operations "centre-valued traces" over $\mathcal{M}$ and $Q\mathcal{M}Q$, respectively. We claim:

there is a monotonously increasing sequence (h_n) of central elements of $\mathcal{M}$ such that, for each $a \in Q\mathcal{M}Q$, (105)

$$\sigma(\mathcal{M}, \mathcal{M}_*)\text{-}\lim_n a^\natural h_n Q = a^{\natural_Q} .$$

Proof. There is an increasing sequence of central projections $\{c_n\}$ such that $c_n Q^\natural$ is invertible in $c_n\mathcal{M}$, for each n, and $\sigma(\mathcal{M}, \mathcal{M}_*)\text{-}\lim_n c_n = c(Q)$.

We define $h_n = (c_n Q^\natural)^{-1}|_{c_n\mathcal{M}} + 0 . c_n^\perp$. Considering the maps $\tau_n(.) = (.)^\natural h_n Q$ over $c_n Q\mathcal{M}Q$, one recognizes all those properties which characterize the $\natural$-operation

$\natural'$ with respect to $c_n Q\mathcal{M}Q$. Thus, by uniqueness of $\natural$-operations, we have $(.)^{\natural'} = \tau_n(.) = (.)^{\natural} h_n Q = (.\, c_n)^{\natural_Q} = c_n(.)^{\natural_Q}$ over $c_n Q\mathcal{M}Q$. The latter, due to the definition of h_n, implies for $a \in Q\mathcal{M}Q$ that

$$\tau_n(a) = a^{\natural} c_n h_n Q = (ac_n)^{\natural} h_n Q = \tau_n(c_n a) = (c_n a)^{\natural_Q} = c_n a^{\natural_Q},$$

i.e. $\tau_n(.) = c_n(.)^{\natural_Q}$ on $Q\mathcal{M}Q$, too. But then,

$$\sigma(\mathcal{M}, \mathcal{M}_*)\text{-}\lim_n \tau_n(a) = \sigma(\mathcal{M}, \mathcal{M}_*)\text{-}\lim_n a^{\natural_Q} c_n = a^{\natural_Q},$$

and (105) is established. In modifying the procedure of the proof on existence of $\&_Q$-maps (Lemma 6-14), we can be assured of the existence of a largest element in $\|\cdot\|$-cl $\{Quau^*Q)^{\natural}\}_{u \in \mathcal{U}}$ (note that, in the finite case, $uu^* = 1$ implies u to be a unitary). Call this element $\pi_Q(a)$. Then, it is not hard to see that $\pi_Q(.)$ shares most of $\&_Q$- resp. $\mathcal{J}_Q$-properties. Especially, due to (105) we have

$$\sigma(\mathcal{M}, \mathcal{M}_*)\text{-}\lim_n \pi_Q(a)\, h_n Q = a^{\&_Q}, \quad \mathcal{J}_Q(a) = \sigma\text{-}\lim_n \pi_Q(a)\, h_n\,. \tag{106}$$

Remembering Lemma 6-16, we can obtain from (106)

$$\pi_Q(a) \leqq \pi_Q(b) \quad \text{for all} \quad Q \in \mathcal{M}^p \quad \text{implies} \quad b \succ a\,, \tag{107}$$

for $a, b \in \mathcal{M}_+$. On the other hand, by definition of $\succ$ on $\mathcal{M}_+$,

$$K(\omega(Q.Q)^{\natural}), b) \geqq K(\omega((Q.Q)^{\natural}), a) \quad \text{for all} \quad \omega \in \mathcal{S} \quad \text{and} \quad \text{for all } Q \in \mathcal{M}^p \tag{108}$$

is necessary for $b \succ a$. By construction of π_Q, however, (108) means

$$\omega(\pi_Q(b)) \geqq \omega(\pi_Q(a)) \quad \text{for all} \quad \omega \in \mathcal{S}\,, \tag{109}$$

so $\pi_Q(b) \geqq \pi_Q(a)$ has to hold, and we have arrived at:

Proposition 6-18. *Let $\mathcal{M}$ be finite, and $\{\pi_Q\}_{Q \in \mathcal{M}^p}$ the above constructed family of centre-valued maps over $\mathcal{M}_+$. Then, for $a, b \in \mathcal{M}_+$, $b \succ a$ iff $\pi_Q(b) \geqq \pi_Q(a)$ for each $Q \in \mathcal{M}^p$.*

It is obvious that π_Q shows up properties as Σ-property, subadditivity, monotonicity, uniform continuity etc. Now, because of $\pi_Q(b) \geqq \pi_Q(a)$ iff $\omega(\pi_Q(b)) \geqq \omega(\pi_Q(a))$ for any *normal* $\omega \in \mathcal{S}$, from (109) and (108), together with linearity and normality of the $\natural$-operation, it follows that:

Lemma 6-19. *In a finite W^*-algebra $\mathcal{M}$, for $a, b \in \mathcal{M}_+$ we have $b \succ a$ if and only if $K(\omega, b) \geqq K(\omega, a)$ for each normal state ω.*

Of distinguished interest is the case where one of the elements to be compared via $\succ$ is a projection:

Lemma 6-20. *Let $\mathcal{M}$ be finite, and $p \in \mathcal{M}^p$, $a \in \mathcal{M}_+$. Then, $p \succ a$ if and only if $o \leqq a \leqq 1$ and $a^{\natural} \leqq p^{\natural}$.*

Proof. Let $q \in \mathcal{M}^p$. We find $z \in \mathcal{Z}^p$ with $qz \succ pz$, $pz^{\perp} \succ qz^{\perp}$. Assume $p_0 \sim p$ with $qz \geqq p_0 z$, $p_0 z^{\perp} \geqq qz^{\perp}$, and suppose $a^{\natural} \leqq p^{\natural}$ with $o \leqq a \leqq 1$. Then, $za^{\natural} \leqq zp_0^{\natural}$ and $z^{\perp}a^{\natural} \leqq z^{\perp}p_0^{\natural}$, and we see, for each $u \in \mathcal{U}$, the relation

$$\begin{aligned} z(quau^*q)^{\natural} &= z(quau^*)^{\natural} \leqq za^{\natural} \leqq zp_0^{\natural} = (zp_0)^{\natural} = (qp_0qz)^{\natural} \\ &= z(qp_0q)^{\natural} \leqq z\pi_q(p_0) = z\pi_q(p)\,, \end{aligned}$$

where $p_0 \sim p$ was used (analogously to Theorem 6-1, 3). The latter gives $z\pi_q(a) \leqq z\pi_q(p)$. On the other hand, because of $a \leqq \mathbf{1}$:

$$z^\perp(quau^*q)^\natural \leqq z^\perp q^\natural = (z^\perp qp_0q)^\natural \leqq z^\perp \pi_q(p_0) = z^\perp \pi_q(p) ,$$

i.e. $z^\perp \pi_q(a) \leqq z^\perp \pi_q(p)$. Taking these relations together, we see $\pi_q(a) \leqq \pi_q(p)$. This has to hold for every projection q, so by Proposition 6-18 $p \succ a$. Suppose $p \succ a$, with $a \in \mathcal{M}_+$. Then, arguing again by Proposition 6-18, $a^\natural = \pi_1(a) \leqq \pi_1(p) = p^\natural$ has to hold, and since $||a|| \leqq 1$ is a trivial consequence of $p \succ a$, the proof is complete. ∎

Let us extend the notion of e_s-functionals (cf. [99, 100, 101]) from type I_n factors (see 1.6., especially (27)), for all finite $\mathcal{M}$ by (see [118]):

Definition 6-21 (generalized e_s-numbers). Let $\mathcal{M}$ be a finite W^*-algebra, with centre $\mathcal{Z}$. We define generalized e_s-functionals $e_z(.)$, with $z \in \mathcal{Z}_+$, by

$$e_z(\omega) = \sup_{\substack{o \leqq b \leqq 1 \\ b^\natural \leqq z}} \omega(b) , \qquad \omega \in \mathcal{M}_+^* .$$

Analogously to Theorem 2-2, we have (cf. [118]):

Theorem 6-21. *Let $\mathcal{M}$ be a finite W^*-algebra, with centre $\mathcal{Z}$. Let $\omega, \varrho \in \mathcal{S}$. Then, $\omega \succ \varrho$ if and only if $e_z(\omega) \leqq e_z(\varrho)$ for each $z \in \mathcal{Z}_+$, with $e_z(.)$ given as in Definition 6-21.*

Proof. By Lemma 6-20 and definition of $(\mathcal{M}_+, \succ)$, for $\nu \in \mathcal{S}$, $p \in \mathcal{M}^p$ we may write

$$K(\nu, p) = \sup_{\substack{p \succ b \\ b \geqq o}} \nu(b) = \sup_{\substack{o \leqq b \leqq 1 \\ b^\natural \leqq p^\natural}} \nu(b) = e_{p^\natural}(\nu) .$$

This, with respect to the first part of Theorem 5-7, gives the implication $e_z(\omega) \leqq e_z(\varrho)$ for all $z \in \mathcal{Z}_+ \Rightarrow \omega \succ \varrho$.

On the other hand, for a general $z \in \mathcal{Z}_+$ the functional $e_z(.)$ is a unitarily invariant, convex, l.s.c. functional on $\mathcal{M}_+^*$, so $\omega \succ \varrho$ implies $e_z(\omega) \leqq e_z(\varrho)$ by Theorem 3-2. ∎

6.8 Examples for $\succ$ on $\mathcal{M}_+$

At the end of this chapter, we shall provide ourselves with some examples for $(\mathcal{M}_+, \succ)$. As it was mentioned in 6.6., we only have to know the "elements" $\mathcal{J}_q(p)$ of the centre-valued matrix (resp. $\pi_q(p)$ in the finite case). By (1')–(4') of Theorem 6-17 and v. Neumann's comparability theorem, we may be content in considering $\{\mathcal{J}_q(p)\}_{q,p}$ for those $p, q \in \mathcal{M}^p$, which are mutually comparable via $\succ$, and, moreover, are finite or properly infinite. In this sense, the following scheme is a complete (and irreducible) one:

Proposition 6-22. *Let $\mathcal{M}$ be a finite W^*-algebra. Then:*

$$\pi_q(p) = \begin{cases} q^\natural & \text{for} \quad p \succ q , \\ p^\natural & \text{for} \quad p \prec q . \end{cases} \qquad p, q \in \mathcal{M} . \tag{110}$$

Let $\mathcal{M}$ be properly infinite. Then:

$$p^{\&_q} = \mathcal{J}_q(p)\, q = \begin{cases} q & \text{for } p \succsim q\,, \\ p_0^{\natural_q} & \text{for } q \text{ finite, and } p \sim p_0 \leqq q \text{ (i.e. } q \succsim p)\,, \\ \text{l.u.b. } \{z \in \mathcal{Z}^p\colon zp \sim zq\}\, q\,, & \text{if} \\ & p \precsim q, \text{ with } q \text{ properly infinite.} \end{cases} \tag{111}$$

The proof is an immediate consequence of the construction of all mappings considered. Now, assume $\mathcal{M}$ is a countably decomposable factor. Then, in case $\mathcal{M}$ is finite, (110) implies

$$\text{type I}_N, \text{II}_1\colon \pi_q(p) \cong \begin{cases} \dim q & \text{for} \quad \dim q \leqq \dim p\,, \\ \dim p & \text{for} \quad \dim q \geqq \dim p\,, \end{cases} \tag{112}$$

where dim (.) indicates the canonical relative dimension, ranging in $\{1, \dots, N\}$ for type I_N, and $[0, 1]$ for type II_1. In case of a type I_N factor, let $a \in \mathcal{M}_+$ and $a_1 \geqq a_2 \geqq \dots \geqq a_N$ be the ordered sequence of eigenvalues of a. Then,

$$\pi_q(a) \cong \sum_{i=1}^{\dim q} a_i\,,$$

by (112), and we may write:

Example 6-23. Let $\mathcal{M}$ be a type I_N factor, $a, b \in \mathcal{M}_+$. Then,

$$a \succsim b \quad \text{iff} \quad \sum_{i=1}^{k} a_i \geqq \sum_{i=1}^{k} b_i \quad \text{for} \quad k = 1, \dots, n\,.$$

Let $\mathcal{M}$ be a factor of type II_1. Let $a \in \mathcal{M}_+$, with the resolution of identity $\{e(\lambda)\}$. Define $f_a(\lambda) = \dim e(\lambda)$, and assume $\lambda = \dim q$. Then, with $g_a(\lambda) = \inf \{\alpha \in \mathsf{R}^1\colon f_a(\alpha) \geqq 1 - \lambda\}$:

$$\pi_q(a) = \pi^\lambda(a)\, 1\,, \quad \text{with} \quad \pi^\lambda(a) = g_a(\lambda)\,[1 - \lambda - f_a(g_a(\lambda))] + \int_{g_a(\lambda)}^{\infty} \beta f_a(d\beta)\,.$$

Example 6-24. Let $\mathcal{M}$ be a type II_1 factor, $a, b \in \mathcal{M}_+$. Then,

$$a \succsim b \quad \text{iff} \quad \pi^\lambda(a) \geqq \pi^\lambda(b) \quad \text{for all} \quad \lambda \in [0, 1]\,.$$

For a countably decomposable factor of type III, things are trivial, for, in the case of $q \neq o$, $a \in \mathcal{M}_+$: $\mathcal{J}_q(a) = \|a\|$, due to (111), is a continuity of $\mathcal{J}_q$ and the Σ-property. Let $\mathcal{M}$ be semi-finite, i.e. type I_∞ and type II_∞. From (111):

$$\mathcal{J}_q(p) \cong \begin{cases} 0 & \text{for } \dim p < \infty, \ \dim q = \infty\,, \\ \dfrac{\dim p}{\dim q} & \text{for } \dim p \leqq \dim q\,, \\ 1 & \text{otherwise}\,. \end{cases}$$

We define $E_t(.) = t\mathcal{J}_q(.)$ for $t = \dim q < \infty$, and $E_\infty(.) = \mathcal{J}_q(.)$ for $\dim q = \infty$ (the latter is possible due to Theorem 6-1, 4.). Let $a \in \mathcal{M}_+$, with the resolution of identity $\{e(\lambda)\}$. We define $h_a(\lambda) = \dim e(\lambda)^\perp$, and set

$$m_a(\lambda) = \inf \{\alpha \in \mathsf{R}^1\colon h_a(\alpha) < \lambda\}\,.$$

From above it follows that $E_\infty(a) = m_a(\infty) = \inf_{h_a(\lambda)<\infty} \lambda$, and

$$E_\lambda(a) = m_a(\lambda)\,[\lambda - h_a(m_a(\lambda))] - \int_{m_a(\lambda)}^{\infty} \beta m_a\,(d\beta) \quad \text{for all} \quad \lambda < \infty\,.$$

Then we have: for $a, b \in \mathcal{M}_+$, $a \succ b$ if and only if

$$E_\lambda(a) \geqq E_\lambda(b) \quad \text{for all} \quad 0 \leqq \lambda \leqq \infty\,.$$

Especially, in the case of a type I_∞, factor $E_n(.)$ is more explicitly: Let

$$\{a_1 > a_2 > a_3 > \cdots\} = \{\lambda \in \operatorname{Spec} a \colon h_a(\lambda) < \infty\},$$

and let n_i be the (finite) multiplicity of a_i. Then, for n with $\sum_{i=1}^{k} n_i \leqq n < \sum_{i=1}^{k+1} n_i$:

$$E_n(a) = \sum_{i \leqq k} n_i a_i + n_{k+1}\left(n - \sum_{i \leqq k} n_i\right),$$

and in the case of $\infty > n > \sum_i n_i$:

$$E_n(a) = \sum_i n_i a_i + \left(n - \sum_i n_i\right) E_\infty(a)\,.$$

References

[1] Abbott, J. C.: Trends in Lattice Theory. Math. Studies, New York 1970.
[2] Alberti, P. M.: Thesis, Leipzig 1973.
[3] Alberti, P. M.: Mixing of states of a type-I-factor. Preprint KMU-HEP-7305, Leipzig 1973.
[4] Alberti, P. M.: A theorem on an interesting property of positive linear forms on W^*-algebras. Preprint KMU-QFT-7, Leipzig 1976.
[5] Alberti, P. M.: On states of a type-II_1-factor. Preprint KMU-QFT-7501, Leipzig 1975.
[6] Alberti, P. M.: A Σ-property for positive linear forms on C^*-algebras. Preprint KMU-QFT-10, Leipzig 1976.
[7] Alberti, P. M.: The Σ-property for positive linear functionals on W^*-algebras and its application to the problem of unitary mixing in the state space. Math. Nachr. **90** (1979), 7.
[8] Alberti, P. M.: On maximally unitarily mixed states on W^*-algebras. Math. Nachr. **91** (1979), 423.
[9] Alberti, P. M.: A theorem on the comparability of projections in finite or countably decomposable infinite W^*-algebras. Wiss. Z. Karl-Marx-Univ. Leipzig **26** (1977), 131.
[10] Alberti, P. M.: Zur unitären Mischung von Zuständen über abzählbar zerlegbaren Faktoren. Wiss. Z. Karl-Marx-Univ. Leipzig **26** (1977), 123.
[11] Alberti, P. M.: Remarks on the Σ-property of positive linear forms on W^*-algebras. Proc. Int. Conf. on Operator Algebras, Ideals, and their Applications in Theoretical Physics. Teubner Texte zur Mathematik, Leipzig 1978, p. 154.
[12] Alberti, P. M.: A characterization of the extremal elements of certain convex sets of stochastic matrices. Z. Wahrscheinlichkeitsth. und verw. Gebiete (subm. 1979).
[13] Alberti, P. M., Uhlmann, A.: The order structure of states in C^*-and W^*-algebras. Proc. Int. Conf. on Operator Algebras, Ideals, and their Application in Theoretical Physics. Teubner Texte zur Mathematik, Leipzig 1978, p. 126.
[14] Alberti, P. M., Uhlmann, A.: A problem concerning positive maps on matrix algebras. Rep. Math. Phys. (to appear).
[15] Alfsen, E. M.: Compact convex sets and boundary integrals. Ergebn. Math. u. Grenzgeb., Bd. 57, Springer-Verlag, Berlin-Heidelberg-New York 1971.
[16] Ananthakrishna, G., Sudarshan, E. C. G., Gorini, V.: Entropy increase for a class of dynamical maps. Rep. Math. Phys. **8** (1975), 25.
[17] Arveson, W.: An invitation to C^*-algebras, Springer-Verlag, New York-Heidelberg-Berlin 1976.
[18] Beckenbach, E. F., Bellmann, R.: Inequalities. Springer-Verlag, Berlin-Göttingen-Heidelberg 1961.
[19] Berberian, S. K.: Baer-$*$-Rings. Springer-Verlag, Berlin-Heidelberg-New York 1972.
[20] Berberian, S. K.: Lectures in functional analysis and operator theory. Springer-Verlag, Berlin-New York-Heidelberg 1976.
[21] Bonnesen, T., Fenchel, W.: Theorie der konvexen Körper. Ergebn. Math. u. Grenzgeb., Bd. 3, Springer-Verlag, Berlin 1934. (Reprint 1974).

[22] Brillouin, L.: Science and information theory. Academic Press, New York 1956.

[23] Calkin, J. W.: Two-sided ideals and convergence in the ring of bounded operators in Hilbert space. Ann. Math. **42** (1941), 839.

[24] Crell, B., de Paly, Th., Uhlmann, A.: H-Theoreme für die Fokker-Planck-Gleichung. Wiss. Z. Karl-Marx-Univ. Leipzig **27** (1978), 229.

[25] Csiszár, I.: Information-type measures of difference of probability distributions and indirect observations. Stud. Sci. Math. Hung. **2** (1967), 299.

[26] Dixmier, J.: Les C^*-algèbres et leurs représentations. Gauthier-Villars, Paris 1969.

[27] Dixmier, J.: Les algèbres d'opérateurs dans l'espace Hilbertien. Gauthier-Villars, Paris 1969.

[28] Dye, H. A., Russo, B.: A note on unitary operators in C^*-algebras. Duke Math. J. **33** (1966), 413.

[29] Emch, G. G.: Algebraic methods in statistical mechanics and quantum field theory. John Wiley & Sons, New York 1972.

[30] Fan, K.: On a theorem concerning eigenvalues of linear transformations. Proc. Nat. Acad. Sci. USA **35** (1949), 652.

[31] Fan, K.: Maximum properties and inequalities for the eigenvalues of completely continuous operators. Proc. Nat. Acad. Sci. USA **37** (1957), 760.

[32] Fano, U.: Description of states in quantum mechanics by density matrix and operator techniques. Rev. Mod. Phys. **29** (1957), 74.

[33] Gaal, S. A.: Linear analysis in representation theory. Springer-Verlag, Berlin-Heidelberg-New York 1973.

[34] Glasman, I. M., Lubitch, J. I.: Finite-dimensional linear analysis (in russian). Mir, Moscow 1969.

[35] Glimm, J., Kadison, R. V.: Unitary operators in C^*-algebras. Pac. J. Math. **10** (1960), 547.

[36] Golden, S.: Lower bounds for the Helmholtz function. Phys. Rev. **137** B (1965), 1127.

[37] Haag, R., Kastler, D.: An algebraic approach to quantum field theory. J. Math. Phys. **5** (1964), 848.

[38] Hardy, G. H., Littlewood, J. E., Pólya, G.: Inequalities. Cambridge Univ. Press. Cambridge 1952.

[39] Horn, A.: On the singular values of a product of completely continuous operators. Proc. Nat. Acad. Sci. USA **36** (1950), 374.

[40] Kadison, R. V.: Isometries of operator algebras. Ann. Math. **54** (1951), 11.

[41] Kadison, R. V.: Lectures on operator algebras. Cargese Lectures on Physics **3** (1967), 41.

[42] Kadison, R. V.: Some analytic methods in the theory of operator algebras. Lecture Notes Math. 140, Springer-Verlag, Berlin-Heidelberg-New York 1968.

[43] Kadison, R. V., Pedersen, G. K.: Equivalence in operator algebras. Math. Scand. **27** (1970), 205.

[44] van Kampen, N.: In "Fundamental Problems in Statistical Mechanics", E. G. D. Cohen (editor), North-Holland, Amsterdam 1962.

[45] van Kampen, N.: Lecture Notes on Statistical Physics, 1965 (unpublished).

[46] Kaplanski, I.: A theorem on rings of operators. Pac. J. Math. **1** (1951), 227.

[47] Krein, M. G., Petunin, Ju. I., Semenov, E. M.: Interpolation of linear operators (in russian), Nauka, Moscow 1978.

[48] Krein, M. G., Gochberg, I. Z.: Introduction to the theory of non-selfadjoint operators (in russian). Nauka, Moscow 1965.

[49] Kullbach, S.: Information theory and statistics. J. Wiley, New York 1951.

[50] Lanford, O. E.: Selected topics in functional analysis. In: Statistical Mechanics and Quantum Field Theory. Gordon & Breach Sci. Publ., New York 1972.

[51] Lassner, G., Lassner, G. A.: On the time evolution of physical systems. Publ. JINR, Dubna, E 2 — 7537, 1973.

[52] Lenard, A.: Generalization of the Golden-Thomson inequality. Indiana Math. J. **21** (1972), 457.

[53] LIDSKII, V. B.: Inequalities for eigenvalues and singular numbers. In: F. R. GANTMACHER, "Theory of Matrices" (in russian). Nauka, Moscow 1966.

[54] LIEB, E.: The classical limit of quantum spin systems. Commun. Math. Phys. **31** (1973), 327.

[55] MARCUS, M., MINC, H.: A survey of matrix theory and matrix inequalities. Nauka, Moscow 1972 (transl. from the Russian).

[56] MEAD, A.: Mixing characters and its applications to irreversible processes in macroscopic systems. J. Chem. Phys. **66** (1977), 459.

[57] MEAD, A. C., SCHLÖGL, F.: Macroscopic aspects of mixing distance in non-equilibrium Thermodynamics. Ann. Phys. **115** (1978), 172.

[58] MEYER, P. A.: Probability and potentials. Mir, Moscow 1973 (transl. from the Russian).

[59] MIRSKI, L.: Results and problems in the theory of doubly stochastic matrices. Z. f. Wahrscheinlichkeitsth. **1** (1963), 319.

[60] VON NEUMANN, J.: Mathematische Grundlagen der Quantenmechanik. Springer-Verlag, Berlin-Heidelberg-New York 1968.

[61] VON NEUMANN, J.: Tomsk. Univ. Rev. **1** (1937), 286, in: Collected works, Vol. IV, p. 205.

[62] NEUMARK, M. A.: Normierte Algebren. VEB Deutscher Verlag der Wissenschaften, Berlin 1959 (transl. from the Russian).

[63] PALMER, T. W.: Characterizations of C^*-algebras I, II. Bull. Amer. Math. Soc. **74** (1968), 538; Trans. Amer. Math. Soc. **148** (1970), 577.

[64] PAULI, W.: Festschrift zum 60. Geburtstag A. Sommerfelds. (Editor P. DEBYE), Hirzel-Verlag, Leipzig 1928.

[65] PIETSCH, A.: Operator Ideals. VEB Deutscher Verlag der Wissenschaften, Berlin 1978/North-Holland Publ. Comp. 1980.

[66] PÓLYA, G.: Remarks on Weyl's note "Inequalities between two kinds of eigenvalues of a linear transformation". Proc. Nat. Acad. Sci. USA **36** (1950), 49.

[67] PUSZ, W., WORONOWICZ, S. L.: Passive states and KMS states for general quantum systems. Commun. Math. Phys. **58** (1978), 273.

[68] RÉNYI, A.: Wahrscheinlichkeitsrechnung. 6th. ed. VEB Deutscher Verlag der Wissenschaften, Berlin 1979.

[69] RINGROSE, J. R.: Lecture notes on von Neumann algebras. Univ. Newcastle upon Tyne, 1967.

[70] RINGROSE, J. R.: Lectures on the trace in finite von Neumann algebras. Lecture Notes Math. 247, Springer-Verlag, Berlin-Heidelberg-New York 1972.

[71] ROTA, G. C.: "Alternierende Verfahren" for general positive operators. Bull. Amer. Math. Soc. **68** (1962), 95.

[72] RUCH, E.: The diagram lattice as structure principle. Theor. Chim. Acta **38** (1975), 167.

[73] RUCH, E., MEAD, A.: The principle of increasing mixing character and some of its consequences. Theor. Chim. Acta **41** (1976), 95.

[74] RUCH, E., SCHRANNER, R., SELIGMAN, TH.: The mixing distance. J. Chem. Phys. **69** (1978), 386.

[75] RUELLE, D.: Statistical Mechanics, Rigorous Results. W. A. Benjamin, Inc., Amsterdam 1969.

[76] RUSSO, B.: Linear mappings of operator algebras. Proc. Amer. Math. Soc. **17** (1966), 1019.

[77] RYFF, J. V.: Orbits of the L_1-functions under doubly stochastic transformations. Trans. Amer. Math. Soc. **117** (1965), 92.

[78] RYFF, J. V.: Measurepreserving transformations and rearrangement. J. Math. Analysis and Appl. **31** (1970), 449.

[79] SAKAI, S.: A characterization of W^*-algebras. Pac. J. Math. **6** (1956), 763.

[80] SAKAI, S.: C^*-algebras and W^*-algebras. Springer-Verlag, Berlin-Heidelberg-New York 1971.

[81] SCHLÖGL, F.: Mixing distance and stability of steady states in statistical non-linear thermodynamics. Z. Physik B **25** (1976), 411.

[82] SCHWARTZ, J. T.: W^*-algebras. Gordon and Breach, New York 1967.
[83] SEGAL, I. E.: A note on the concept of entropy. J. Math. Mech. **9** (1960), 623.
[84] SIMON, B.: Identifying the classical limit of a quantum spin system. Princeton University preprint, 1979.
[85] STØRMER, E.: Positive linear maps of operator algebras. Acta Math. **110** (1963), 233.
[86] STØRMER, E.: Positive linear maps of C^*-algebras. Springer-Verlag, Berlin-Heidelberg-New York 1974.
[87] TAKESAKI, M.: On the conjugate space of operator algebras. Tohoku Math. J. **10** (1958), 194.
[88] TAKESAKI, M.: On the singularity of positive linear functionals on operator algebras. Proc. Japan. Acad. **35** (1959), 365.
[89] THIRRING, W.: Vorlesungen über Mathematische Physik, Bd. IV. Springer-Verlag, Berlin-Heidelberg-New York (to appear). (Manuscript version: Wien 1975.)
[90] THOMSON, C. J.: Inequality with application in Statistical Mechanics. J. Math. Phys. **6** (1965), 1812.
[91] THOMSON, C.: Inequalities and partial order on matrix spaces. Indiana Math. J. **21** (1972), 469.
[92] UHLMANN, A.: Observables. Acta Univ. Wratislaviensis, Proc. Winter School f. Theor. Phys., Karpacz, Vol. II., Wrocław 1964.
[93] UHLMANN, A.: On the Shannon entropy and related functionals on convex sets. Rep. Math. Phys. **1** (1970), 147.
[94] UHLMANN, A.: Sätze über Dichtematrizen. Wiss. Z. Karl-Marx-Univ. Leipzig **20** (1971), 633.
[95] UHLMANN, A.: Endlichdimensionale Dichtematrizen I. Wiss. Z. Karl-Marx-Univ. Leipzig **21** (1972), 421.
[96] UHLMANN, A.: Endlichdimensionale Dichtematrizen II. Wiss. Z. Karl-Marx-Univ. Leipzig **22** (1973), 139.
[97] UHLMANN, A.: Unitarily invariant convex functions on the state space of type I and type III von Neumann algebras. Rep. Math. Phys. **7** (1975), 449.
[98] UHLMANN, A.: Zur Beschreibung irreversibler Quantenprozesse. Sitzungsber. Akad. Wiss. DDR **14**/N, Akademie-Verlag, Berlin 1976.
[99] UHLMANN, A.: Markov master equation and the behaviour of some entropy-like quantities. Rostocker Phys. Manuskripte **2** (1977).
[100] UHLMANN, A.: The order structure of states. Proc. Intern. Sympos. on selected topics in Statistical Mechanics. D 17-11490, Dubna, USSR, 1978.
[101] UHLMANN, A.: Ordnungsstrukturen im Raum der gemischten Zustände und die H-Theoreme von Felderhof-van Kampen. Wiss. Z. Karl-Marx-Univ. Leipzig **27** (1978), 213.
[102] UHLMANN, A.: Remarks on the relation between quantum and classical entropy as proposed by A. Wehrl. Rep. Math. Phys. (to appear).
[103] UMEGAKI, H.: Kodai Math. Sem. Rep. **14** (1962), 59.
[104] WEHRL, A.: How chaotic is a state of a quantum system. Rep. Math. Phys. **6** (1974), 15.
[105] WEHRL, A.: On some inequalities involving eigenvalues of linear operators. Acta Univ. Wratislaviensis, 1973.
[106] WEHRL, A.: Convex and concave traces. Acta Phys. Austriaca **37** (1973), 361.
[107] WEHRL, A.: Dichtematrizen in Quantensystemen bei unendlicher Temperatur. Acta Phys. Austriaca **41** (1975), 197.
[108] WEHRL, A.: Remarks on A-entropy. Rep. Math. Phys. **12** (1977), 385.
[109] WEHRL, A.: General properties of entropy. Rev. Mod. Phys. **50** (1978), 81.
[110] WEHRL, A., YOURGRAU, W.: On the increase of entropy in the Carleman model. Physics Letters **72** A (1979), 13.
[111] WEYL, H.: Inequalities between two kinds of eigenvalues of a linear transformation. Proc. Nat. Acad. Sci. USA **35** (1949), 408.
[112] WORONOWICZ, S. L.: Positive maps of low dimensional matrix algebras. Rep. Math. Phys. **10** (1976), 165.

[113] Crell, B., Uhlmann, A.: An example of a non-linear evolution equation showing "chaos-enhancement". Letters Math. Phys. **3** (1979), 463–465.
[114] Carleman, T.: Sur la théorie de l'équation intégrodifférentielle de Boltzmann. Acta Math. **60** (1933), 91.
[115] Simons, S.: Is the solution of the Boltzmann equation positive? Phys. Lett. **69** A (1978), 239.
[116] Alberti, P. M., Uhlmann, A.: Existence and density theorems for stochastic maps on commutative C^*-algebras. Math. Nachr. **97** (1980), 279–295.
[117] Alberti, P. M., Uhlmann, A.: A note on stochastic dynamics in the state space of commutative C^*-algebras. Leipzig-Preprint, KMU-QFT-5 (1979); J. Math. Phys. (in print).
[118] Alberti, P. M.: Thesis B, Leipzig 1979.

Symbols

(first occurrence in a given context)

Key words

(first appearance and first appearance with changed definition)

Mathematics and Its Applications

1. Willem Kuyk, *Complementarity in Mathematics, A First Introduction to the Foundations of Mathematics and Its History.* 1977.
2. Peter H. Sellers, *Combinatorial Complexes, A Mathematical Theory of Algorithms.* 1979.
3. Jacques Chaillou, *Hyperbolic Differential Polynomials and their Singular Perturbations.* 1979.
4. Svtopluk Fučik, *Solvability of Nonlinear Equations and Boundary Value Problems.* 1980.
5. Willard L. Miranker, *Numerical Methods for Stiff Equations and Singular Perturbation Problems.* 1980.
6. P. M. Cohn, *Universal Algebra.* 1981.
7. Vasile I. Istrăţescu, *Fixed Point Theory, An Introduction.* 1981.
8. Antoni Wawrzyńczyk, *Group Representation and Special Functions.* 1982 (forthcoming).
9. Peter M. Alberti and Armin Uhlmann, *Stochasticity and Partial Order. Doubly Stochastic Maps and Unitary Mixing.* 1982.